T0222293

Science and Fiction

Science and Fiction—A Springer Series

This collection of entertaining and thought-provoking books will appeal equally to science buffs, scientists and science-fiction fans. It was born out of the recognition that scientific discovery and the creation of plausible fictional scenarios are often two sides of the same coin. Each relies on an understanding of the way the world works, coupled with the imaginative ability to invent new or alternative explanations—and even other worlds. Authored by practicing scientists as well as writers of hard science fiction, these books explore and exploit the borderlands between accepted science and its fictional counterpart. Uncovering mutual influences, promoting fruitful interaction, narrating and analyzing fictional scenarios, together they serve as a reaction vessel for inspired new ideas in science, technology, and beyond.

Whether fiction, fact, or forever undecidable: the Springer Series "Science and Fiction" intends to go where no one has gone before!

Its largely non-technical books take several different approaches. Journey with their authors as they

- Indulge in science speculation—describing intriguing, plausible yet unproven ideas;
- Exploit science fiction for educational purposes and as a means of promoting critical thinking;
- Explore the interplay of science and science fiction—throughout the history of the genre and looking ahead;
- Delve into related topics including, but not limited to: science as a creative process, the limits of science, interplay of literature and knowledge;
- Tell fictional short stories built around well-defined scientific ideas, with a supplement summarizing the science underlying the plot.

Readers can look forward to a broad range of topics, as intriguing as they are important. Here are just a few by way of illustration:

- Time travel, superluminal travel, wormholes, teleportation
- Extraterrestrial intelligence and alien civilizations
- Artificial intelligence, planetary brains, the universe as a computer, simulated worlds
- Non-anthropocentric viewpoints
- Synthetic biology, genetic engineering, developing nanotechnologies
- Eco/infrastructure/meteorite-impact disaster scenarios
- Future scenarios, transhumanism, posthumanism, intelligence explosion
- Virtual worlds, cyberspace dramas
- Consciousness and mind manipulation

For further volumes:
http://www.springer.com/series/11657

Jayant V. Narlikar

The Return of Vaman— A Scientific Novel

 Springer

Jayant V. Narlikar
Inter-University Centre
for Astronomy and Astrophysics
Pune
India

ISSN 2197-1188   ISSN 2197-1196 (electronic)
Science and Fiction
ISBN 978-3-319-16428-1   ISBN 978-3-319-16429-8 (eBook)
DOI 10.1007/978-3-319-16429-8
Springer Cham Heidelberg New York Dordrecht London

Printed on acid-free paper

Springer International Publishing is part of Springer Science+Business Media (www.springer.com)

Preface

It gives me great pleasure to present this sample of my science fiction work to the international reader. So far my sci-fi stories and novels have only been published within India where they have received a warm welcome. However, this happens to be the first occasion when a novel and a short story are being jointly published by Springer.

Since this is the first time that my science fiction is being projected abroad, the publisher made the very pertinent suggestion that I also write an introductory article describing my association with this form of literature. It is all the more pertinent because by profession I am a scientist working in the field of astrophysics. This exercise includes a brief description of my background and why and how I got into writing sci fi. Knowing full well that other sci fi writers may have different viewpoints, I felt that I should stick to my own views in a personalized autobiographical statement.

Life in India is inextricably mixed with Indian mythology and this shows up in the two examples of sci fi presented here. This is peculiar to India, and one may wonder how ancient myths can be combined with futuristic ideas that arise in sci fi. Thus Ganesha and Vaman are part of those myths but, through various rituals, they have become part of our modern world of jets, computers and cell phones. So why not take a step into the future and integrate them into science fiction?

Anyway, since my views *in extenso* are given in my introductory article, I will be brief here. It only remains for me to thank the Springer publication staff, and in particular Chris Caron, for a very helpful interaction. Their help and advice are much appreciated.

Inter-University Centre   Jayant V. Narlikar
for Astronomy and Astrophysics
Pune, India
January, 2015

Contents

Contents

Part I

A Short Story: Ganesha (1975)

The Rare Idol of Ganesha

As I stepped off the bus outside the Oval, I had a premonition that I was going to witness something unusual. Today, looking back after the events, I see no reason to account for that premonition, but then aren't premonitions, by definition, unaccountable? So far as I can see, the only thing unusual was that I had found time to watch a test match live; and this was fully accountable. As I presented my complimentary pass at the gate I fingered the note which accompanied it:

14 August 2005

Dear John,

I sincerely request you to watch my performance in what is going to be my final test appearance. Hope you can make it!

Regards,

Sincerely yours,

Pramod Rangnekar

Pramod and I were Cambridge blues in the team which won the Varsity match in '89. Later, Pramod became a professional, rising to great heights as his performances against the West Indies, Australia and the M.C.C. have testified. I, regrettably, gave up cricket altogether, such were the demands of my work as an Indologist and (now) as a museum curator. Indeed it came as a shock to me that I was visiting a cricket ground after 15 years!

Today was the second day's play and I had chosen this day because India was going to field and I would be able to watch Pramod in action. Yesterday, on a perfect wicket in bright sunshine India was expected to amass a huge score. However, the Indian batsmen did not fulfill these expectations and England was left to face a relatively modest first innings score of 308.

The second day's play began quietly with no hint of what was to come. In one hour the England openers Willis and Jones put up a score of 40 for no loss. After the drinks, the Indian captain Bhandari called in Rangnekar to bowl—and the unusual chain of events was set in motion.

Pramod was greeted with great applause which had an element of sympathy about it. For, as all followers of the game knew, this was his last test. Indeed, during this series his performance had been indifferent. The old fire and magic had gone from his bowling, which could now be easily 'read' by the England batsmen. There had been an increasing demand that Pramod be dropped from the team. Nevertheless, the selectors once again plumped for experience, rather than quality and gave him this last chance. Would he live up to their expectations?

The first hint of the unusual came when Pramod prepared to bowl, "Right arm, over the wicket?" asked umpire Coates, who was familiar with his style of bowling. "No," said Pramod to the umpire's surprise. "Left arm over the wicket."

Pramod had never bowled with his left arm. Even Bhandari was puzzled and wanted to discourage Pramod. However, Pramod persisted. "OK, I will allow the old … one over of this nonsense," muttered Bhandari to himself.

By now the commentators on radio and TV had learnt of Pramod's intention and had started commenting on it. How could a right-handed bowler suddenly decide to bowl left-handed? And that too in a test match when his side badly needed a wicket? This was against all precedents.

But then everything that followed was going to be against any precedent.

With his first ball, which was remarkably accurate, Pramod knocked off opener Willis's leg stump. As a confounded Willis made his way back to the pavilion he warned the new batsman, "Take care! The joker sent down a funny one to me."

The warning was to be of no avail. Pramod's left-handed bowling was completely unexpected and neither No. 3, nor any of his successors could make any sense of it. From 40 for no wicket, England was bundled out for a total of a mere 78 runs.

As I munched my sandwich during the lunch interval, I pondered over the remarkable transformation which had come over the game in such a short time—a transformation which perhaps distinguishes cricket from any other game. Would England recover in the second innings as they followed on 230 runs behind? The old gentleman sitting next to me wondered moodily whether Rangnekar would overtake Jim Laker's record of 19 wickets in a test match. He went on describing that eventful match which he had seen in his youth half a century ago.

Yes, this is what happened in the second innings when England collapsed once again for 45 runs, their lowest ever score against India. And Pramod had a tally of 20 wickets.

Then this remarkable match was followed by another remarkable event. As the last wicket fell Pramod ran towards the pavilion and even before the jubilant Indian spectators, excited newsmen and TV crew could get anywhere near, he was whisked away in a waiting car.

Where did Pramod go? Nobody knew. The manager of the Indian team, the police and the newspapermen began a frantic search. The next day, an unknown person with an Indian accent telephoned a Fleet Street newspaper office and conveyed a message from the missing player:

"I am safe; don't worry, I will return within 24 hours."

Was this a hoax or a genuine message? By way of authentication the caller told the police a location in Croydon where Rangnekar's shirt would be found. Sure enough the police located and identified the shirt.

Meanwhile, the newspapers had a field day. 'An Indian Rope-trick?' 'Superb bowling or Eastern hypnosis?' 'Rangnekar out-Lakers Jim Laker', blared some of the tabloid headlines. Even the *Times* felt driven to writing an editorial eulogizing Rangnekar's achievement but confessing to being puzzled at developments during and after the game. The Wisden promptly added another record to its annals of cricket history.

The next day Pramod was found at the Bow Street police station. But what an anticlimax! He did not remember a single thing about the test match or what happened afterwards. In all other respects his brain was sound and he ridiculed any suggestions that he had played such a major role in the test match.

"I could not get the wicket of a schoolboy if I bowled left-handed," he said modestly—and there was a ring of truth in his voice.

13 December 2005 is a date I will never forget. I had finished my breakfast and was about to leave for an important appointment when the phone rang:

"John, it's for you. The caller won't identify himself but says it's important," Ann said. I cursed inwardly—I would certainly be late for my appointment now.

"Yes? John Armstrong speaking," I tried to be as polite as possible.

"Good morning, John! You will be surprised to hear from me. This is Ajit calling—Ajit Singh."

Ajit Singh! After so many years! Annoyance gave way to surprise as I continued to listen. "May I see you tonight? It is very important. About eight thirty?" He seemed to be dictating all the arrangements. I asserted myself, "Come for dinner. Ann is threatening to poison me with her curry. We will both be her victims."

"Sure, thanks," said Ajit. As he was about to hang up he seemed to remember something. He added, "And John—I hope you and Ann will not mind if I eat with my hands instead of with knife and fork."

Why this reference to knife and fork? Before I could ask him if he was serious, Ajit rang off.

Ann was only too happy to try her hand at a curry. She also decided to experiment with some Indian sweets. I left her with her cookery cards and hastened for the train. But throughout the day my thoughts were on Ajit and on our forthcoming encounter. What was he going to tell me?

Pramod and Ajit were fellow undergraduates with me at Cambridge. We had rooms on the same staircase of the college. Pramod and I shared an enthusiasm for cricket and by the end of our first summer we were both picked for the University Eleven. With Ajit I had a different type of bond. We both used to hold long discussions, lasting sometimes into the early hours of the morning, on Indian philosophy. Indeed it was these discussions which really shaped my career as an Indologist. Ajit, however, was a physicist. After taking the third part of the Mathematical Tripos, where he won the Mayhew Prize, he elected to do physics. Here also he distinguished himself. I left Cambridge after three years but he continued a research career at the Cavendish. Off and on we had met and corresponded; I do recall writing to him when he won the Smith Prize. But later our contact was less frequent. I had been on several archaeological expeditions in the Indian subcontinent before settling down to my present museum curatorship in London.

Ajit had been a loner all along. I doubt if he ever had any friend apart from me. When we last met, which was five years ago, Ajit had given up his college Fellowship and joined a research establishment in England. I believe, though he never mentioned it, his work was of a highly classified nature.

Was he going to tell me something of it tonight?

Exactly at 8.30 pm the door bell rang. I had no trouble recognising Ajit. He had become leaner and had a few grey hairs. But there was another subtle change in him which I could sense—I can testify to that even today after all these events. To be honest, however, I must also record that at that moment of our meeting I was not able to pinpoint what exactly was different about him.

His manner of speaking soon put my mind at rest. So far as his attitude towards me was concerned, it hadn't changed a bit.

At the meal, which was served in Indian fashion in *thalis* (another of Ann's attempts at artistic verisimilitude!), Ajit was reticent, leaving aside the usual small talk. But this did not surprise me as Ajit was never a sparkling dinner-table conversationalist. What did surprise me was his manner of eating. Out of deference to his whimsical suggestion we had all dispensed with knife and fork in favour of fingers. But Ajit's way of eating with fingers showed the same awkwardness that a Westerner exhibits when he attempts to eat Indian food with fingers. Ann and I commented on this. But Ajit had an explanation: "Living in the decadent West for so many years, I have lost the knack of eating with fingers." The explanation seemed to satisfy Ann, but I had my doubts.

My doubts about Ajit's unusual behaviour were reinforced towards the end of the meal when my seven-year-old son Ken came with a book.

"Uncle, you sent me this book on my last birthday but you forgot to sign it. Would you please do it now?"

This was a fact. Last year a book had arrived for Ken from Ajit's lab. It carried the inscription 'To Ken, on his seventh birthday' in what I knew to be Ajit's handwriting. Ajit in his peculiar way had remembered Ken's birthday but had forgotten to sign his name!

Ajit took the book and glanced at it in a cursory fashion. Then he shook his head and returned it to Ken.

"I am sorry, Ken! My eyes are hurting me today so I can't sign this right now."

"Come on! You don't need to exert your eyes to sign your own name," I protested on Ken's behalf.

"But my doctor has expressly forbidden me to read or write anything in my present condition. As a compromise, Ken, I will bring you another book soon when I am well and I will sign both of them."

Ajit's tone had an air of finality; so Ken and I did not press further. Ken appeared satisfied with the offer of another present—he had already developed a liking for books. But I found Ajit's response highly uncharacteristic of him.

"Now Ajit, perhaps you can tell me why you came here tonight." My suppressed curiosity finally burst out as I pointed him to an armchair in my study and offered him a glass of port. We were alone now and I expected something momentous from him.

"Take it easy!" Ajit had a relaxed smile on his face. He slowly took out a packet from his briefcase and opened it carefully.

It was a beautiful idol of the dancing *Ganesha*, the elephant god of the Hindus. (The elephant god has the head of an elephant and its idol is usually in a sitting posture with legs crossed as in the Buddha's sitting statues. This particular idol showed the elephant god in a dancing pose, which is not so common.) I recognised it immediately, for a similar idol existed in my museum in the British India Section. It belonged to the Maratha rulers, the Peshwas who controlled most of India before the British became dominant. *Ganesha* was one of the important deities of the Peshwas and this particular idol in my museum had been recovered from their palace, the *Shaniwarvada*, when Elphinstone's army marched into Pune in 1818. How it finally made its way to this museum is a long story. My immediate reaction was to ask how Ajit managed to get a replica of this valuable piece.

"Look carefully! Is it really a replica?" Ajit had a provocative smile on his face.

I subjected the piece to the many visual tests of authenticity that I knew. Yes, so far as I could tell this piece was made by the same craftsman who had made the idol I had in the museum. Then suddenly, I noticed one glaring difference: how could I miss it in the first place?

The trunk of the elephant head was turned to the right instead of to the left as with most idols of *Ganesha*.

This particular aspect not only distinguished the idol in my hand from that in the museum but it also made it far more valuable because of its rarity. I explained this to Ajit.

"Indeed? I would like to see them side by side for comparison." Ajit seemed more amused than surprised. He continued, "May I present your museum with this piece since you find it so valuable?"

I thanked him for this generous gift and promised him a properly worded formal letter of gratitude from the trustees of the museum. But I could not contain my curiosity and asked, "What is the history behind this piece? How did you come by it?"

"All in good time: but I am happy to see you so surprised. Let me now ask you another question, John. You know me well. What do you think is a distinguishing mark of my body?" I was surprised by this sudden change of subject. But of course I knew the answer.

"Your left thumb is about half an inch smaller than your right thumb."

"Can you swear to it?"

"Of course!"

Ajit opened out both his hands in front of me. Yes, one thumb was shorter than the other. But I realised with a shock that it was the right thumb that was short.

I must have passed out with the shock, for when I came to I found Ajit gazing anxiously at me with a glass of brandy in hand. "Are you OK?" he asked.

"Just who the devil are you?" I asked aggressively. I was conscious of a chagrin at having displayed a weakness earlier and was trying to compensate for it.

"I am Ajit and none other—only I am slightly changed." Ajit picked up my right hand and held it to his chest. His heart was beating on the right.

A crazy but connected picture began to form in my mind. I made Ajit stand in front of a mirror in my study, and I got the answer to a nagging thought which had been with me all evening since Ajit's arrival. In some subtle way he had appeared different. Now I could see that the image staring at me from the mirror was more familiar to me than the live figure I was holding by his shoulders. Had Ajit somehow managed to convert himself into his mirror image?

I recalled the fantastic left-handed bowling of Pramod. Was it real Pramod or was it his image? Surely it was no illusion, because I had not been the only person to watch him perform. Even instruments like cameras and TV had conveyed the same effect. Was the *Ganesha* idol also an image of the real one?

I went to the desk to feel it. It was as solid and real as Ajit grinning in front of me.

"I am sorry to have shocked you—but there was no other way of convincing you of the fantastic discovery I have made. I will begin, as they say in books, at the beginning ..."

"Before you do that, please tell me one thing. Am I right in supposing that you are behind Pramod's mysterious performance?"

"Of course!" said Ajit and he began his story which I give below in his own words as far as possible.

You might recall that about five years ago I had left Cambridge to take up some classified research work in a government lab. I had brought with me expertise in fundamental physics and electronics along with an innocent enthusiasm for work. The latter I quickly shed aside as I found that, instead of research, the main emphasis was on desk work, sycophancy and politicking.

With growing disillusionment I began to cut myself off from my colleagues who seemed only interested in idle gossip. I made sure that I did the work assigned to me promptly. As this was very little, I found myself with a lot of spare time to do my own thinking and research. Knowing my introvert nature, my colleagues and superiors also left me alone.

I had long been toying with a curious concept which came to my mind when I studied Einstein's general theory of relativity at Cambridge. As you may be unfamiliar with this theory, let me describe its salient points which were of use to me.

Einstein introduced the idea that gravitation modifies the geometry of space and time. We are familiar with the geometry of Euclid which seems to serve us well in our daily life. Yet, nearly a century and a half ago mathematicians had begun to realise that Euclid's geometry need not be the only logically consistent geometry. Non-Euclidean geometries based on rules different from Euclid's axioms could be thought of. However, Einstein in 1915 was the first scientist to employ these abstract ideas in a physical theory. He argued that massive gravitating objects have non-Euclidean geometries around them—and he gave equations to describe these. Some of his predictions were verified in the second half of the last century.

Take, for instance, light rays which are supposed to travel in straight lines. The meaning and criterion of a straight line are different in different geometries. Near the Sun, its strong gravity will modify the geometry significantly so that, if a light ray passes close to the solar limb, its track will be different from what it would have been in the absence of the Sun. Such differences, although small, were measured and they confirmed the predictions of general relativity.

In the jargon of relativity, we say that the geometry of spacetime is 'curved' instead of 'flat' when gravitation is present. A two-dimensional flat creature moving on the surface of a sphere is conscious of the curvature of the surface. Imagine a similar curvature in higher dimensions—it is difficult conceptually, but easy mathematically!

Now I will introduce another concept into the picture—that of twist. Have you heard of the Möbius strip? If you wear your belt with one twist you will get this strip. It has many peculiarities. For example, unlike the original belt which had two surfaces, this has only one. If you cut an ordinarily tied belt along a line at half width, you will get two separate belts. Try doing the same with the Möbius strip and you will be in for a surprise.

Now imagine our flat creature crawling in this single surface of the Möbius strip. Suppose he has only one hand say, the left hand. However, if he makes one round of the strip he will find that the one hand he has is his right hand! If you do not believe it try it on a paper strip. To us, observing from the vantage point in a three-dimensional space the creature has undergone a half rotation round an axis passing through his body from head to foot. But the creature is not conscious of this. To him, in his limited two-dimensional perception, this appears as a reflection.

Now imagine a similar twist in our four-dimensional spacetime. Like the creature, we will not be conscious of it, except through similar effects. By going through it, we will appear as our mirror reflections, whereas in fact we are being turned round in high dimensions and showing our 'other side'.

Can we produce such twists in spacetime? It is here that I departed from Einstein's theory and had my own conjecture. I expected that the property of spin found in subatomic elementary particles could generate twists in space. I had the mathematical ideas fully worked out in Cambridge. To put these into practice became possible in my present establishment.

To generate a substantial twist I had to make a beam of elementary particles with spins not randomly oriented but well-coordinated. This is sooner said than done and I will not bore you with the intellectual convulsions I went through. I will only say that I succeeded for the first time about six months ago.

Thanks to the prevailing atmosphere in my lab I was able to construct equipment with almost no interference from anyone. I had erected a small cabin around it and put notices 'top secret', 'dangerous' and 'do not enter' on it. Nobody bothered to ask what I was doing, so long as I kept within my funds. In a highly bureaucratic system it is possible to wangle things if one is clever enough. The alternative was to submit a proposal, then have it evaluated and most likely rejected by a committee which would rely on so-called expert opinion from people long past the stage of active research.

My first experiment was with my wrist watch. Apart from a mirror reflection I wanted to see whether its mechanism would survive the transformation.

It did. This was important because my next step involved live objects. I experimented on insects, butterflies, guinea pigs, etc. When I found that even living objects survive the experiments, I decided to take the final step.

Knowing the many possible outcomes of such a step I wrote all the details and kept them in a safe place. I set up video and tape equipment to 'observe' and 'hear' the outcome as I submitted myself to the reflecting machine.

My experience as I went through the beam produced by the machine was surprisingly normal. Never was I conscious of being twisted or contorted. There was no discomfort as I walked round the beam. I spoke out whatever I felt and this was duly taped.

As I came out I found that I had indeed been transformed. Not only that, my dress, wristwatch, pen, everything on me went through that change. My brain had also been transformed so that I found it difficult to read normal writing. All operations which distinguished between left and right were confusing. I had to think which way to turn the screwed top of a bottle in order to open it—for my new instinct dictated the wrong way. But physically I was fit and felt my left hand to be much stronger and versatile than my right hand.

Then, to complete my experiment I went through the beam again. As expected I regained my usual form as I emerged, but with one important fact which I had anticipated.

My brain retained no memory of my reflected state!

It was only through the evidence recorded during my transformation by the various instruments that I could convince myself that it had in fact happened. I looked at notes made by me in the reflected state. I could not read them until I saw them reflected in a mirror!

This erasing of memory of what happened in the reflected state is an unsatisfactory feature of my experiment which I have not so far been able to rectify. When I change myself back to my usual form I will have totally lost all memory of my encounter with you tonight!

As I listened to Ajit's weird tale I had the feeling that all this was not real—but a mixture of Lewis Carroll, H.G. Wells, and the Arabian Nights. But I was looking at the living proof quietly sipping port in front of me. To set any remaining doubts to rest, I asked Ajit the question which had been bothering me:

"Is this *Ganesha* also a reflection?"

"Why don't you verify it yourself? You live right above the museum." Ajit's suggestion was a practical one.

I took a bunch of keys and we both went down to the British India Section. By the time I reached for the cabinet where the *Ganesha* was supposed to be locked in, I knew what I would see.

The cabinet was empty!

"So I was not such a generous donor after all!" quipped Ajit, as I returned to my study after placing Ajit's 'gift' in the empty cabinet. He must have somehow pinched the original and subjected it to his infernal experiment.

"What about Pramod's performance?" I asked. Surely, all that I had learnt so far shed considerable light on the mystery.

"Pramod came to see me on the eve of the test match. He was very depressed. He knew that he was past his prime as a test match bowler and that his inclusion in the final test was not purely on merit. It was something he then said that gave me a daring idea. 'There are no surprises left in my bowling,' was what he moaned about.

"Suppose I turned him into his reflection? I thought he would bowl as a left-hander but not as an ordinary left-handed bowler would. All his actions would be that of a right-hander reflected in the mirror. In any case, none of the batsmen expected him to bowl like that.

"I drugged his coffee, and while he was unconscious, subjected him to my experiment. Taking him to the lab in spite of the tight security was no problem. I had discovered the loopholes in the security system long before. After the experiment I left him on his hotel bed.

"Early next morning, I had a frantic phone call from him. He was hysterical—he felt weak, could not read, found letters inside out … He wanted to know if he had eaten something at my place last night that caused this trouble. He was scared to call the doctor lest he was declared unfit for the match.

"I rushed to his room to reassure him. His right hand had gone weak and he could not bowl. What about his left hand? Surprisingly it was in good condition and I suggested he bowled with it. He found the idea ludicrous—but the more he swung his hand the more reasonable it appeared to him. He suggested that he should have net practice. As his team-mates were still in their beds, I offered to take him to the practice enclosure. This turned out much better since his new-found prowess could be kept secret from everybody until the crucial moment. You know the rest," Ajit concluded.

"Was it you who spirited him away after the match?" I asked.

"Yes. And it was I who telephoned to give the message to the newspaper. I had kept him in my flat for a couple of days. When he recovered sufficiently I transformed him back to normal and delivered him at Bow Street. Of course he had totally forgotten all his traumatic experiences. "I felt it unwise to tell him the truth."

As a good scientific theory can explain many phenomena so were all my mysteries resolved by this remarkable discovery of Ajit. I could also see why he did not want to eat with knife and fork. Ann and I would have detected his awkwardness. As it was, we did comment on his difficulty in eating with fingers but he had a reasonable explanation for it. What really caught him by surprise was Ken's request for an autograph. Even signing one's name can be very difficult if your brain insists on projecting all letters the wrong way round!

"Ajit, you must publish all your findings at once. You are sure to get the Nobel Prize." This was my advice as a layman.

"No, not yet, John," Ajit replied. "You know I am a perfectionist and I find the loss of memory a grave defect in my work. Until I remove this defect I am not prepared to announce my discovery to the world."

"But Ajit, let me offer you some practical advice. You are playing with unknown laws of nature. That you have achieved success so far does not guarantee that you will succeed again. Wouldn't it be wise to keep a clear record of all you have done in a safe place?"

"Of course, I have done that. After reading my account any scientifically competent group can repeat my experiment. As regards your statement about future success, I do not deny it. But I am in the process of modifying my experiment which I feel will soon remove the final blemish. Indeed I would not have revealed to you my progress so far, but for my impish desire to surprise the only real friend I have."

I tried to argue with him; but as I had feared, once Ajit's mind was made up it was impossible to change it.

A few months later I received a phone call from Ajit's lab. I was hurriedly summoned to see the Director.

With ill forebodings, I knocked on the door. In his office were sitting the Director, a doctor in a white coat, a non-descript man and Ajit. I breathed a sigh of relief—I had feared to see him dead.

But my relief was only short-lived. Ajit did not recognize me. Indeed, as the doctor explained, Ajit was suffering from a totally irreversible amnesia. It was only because they had found my name and telephone number in his office that his lab could contact me.

"Has he left any written records of what he was doing?" I asked cautiously. I remembered with chagrin that I had forgotten to ask Ajit in what 'safe place' he had kept his records.

"If he did, we have unfortunately no means of knowing," sighed the Director. "You see, whatever experiment he was doing blew up and shattered everything in his room."

"He is lucky to be alive," commented the doctor. What an ironical choice of words! For a genius like Ajit this loss of memory was worse than death.

"What about his house?" I asked, hoping that there might be something there.

"A typical bachelor's mess," commented the non-descript man. "We searched his flat with a fine-toothcomb. There is nothing there. Indeed we called you here to ask whether you could throw any light on the matter."

"Sorry, I can't help you there. I am afraid Ajit, though a good friend of mine, never considered me educated enough to share his scientific confidences."

As I drove home, I wondered whether truth would have been more convincing than lie. I decided in the negative. After all, my unscientific description would have sounded too fantastic to be credible.

Even today I find it fantastic but not incredible! For, you see, I have a concrete proof lying in my museum. It is the rare idol of *Ganesha*.

Part II

The Novel:
The Return of Vaman (1986)

The Container

1 The Transfer

'The seat belt sign has been switched off … but it is advisable to keep the seat belts fastened during flight …'

The announcement was hardly over before the passenger in seat 59A sprang up and rushed to the front of the plane. The stewardess was startled by this burst of activity … but soon ceased worrying. The passenger entered the toilet.

'Took him for a highjacker, did you, Sheela?' smiled Jamshed twirling a moustache that would have done justice to Air India's Maharajah.

'Well, one never can tell these days, can one?' observed Sheela pointing to 60A. 'Look over there. That one is up to no good.'

Without turning his head Jamshed managed to look.

'You are nervous these days, Sheela. Won't be surprised if you start doubting our captain's intentions next! … 60A has had one too many, that's all.'

60A was a burly European. When all the announcements were over he too rose and made his way to the toilet. Unbuttoned shirt and denims … his hefty, masculine body seemed to burst out of his clothes. But at that moment all of it was shaking and tottering.

By now *Emperor Vikramaditya* was well set on its flight path at 30,000 feet. That the flight was steady could be seen from the unruffled levels of drink in the glasses resting on the passengers' tables. But 60A was nevertheless finding his progress towards the toilet difficult.

He finally reached his destination and began struggling with the door handle. Jamshed tapped him gently.

'Sir, this toilet is occupied. There are more toilets on the lower deck in case you wish …'

'*Danke! Nein!* … Thank you … Will wait here.'

Looking down at the narrow, winding staircase to the lower deck, Jamshed could appreciate 60A's point of view. He was on the point of offering his stewards's seat to the passenger when the toilet door opened. 59A emerged and, casting a look of disapproval at 60A, made his way back to the seat. 60A rushed in.

'Come, time to serve the snacks', Jamshed activated Sheela.

When 60A came out, both Jamshed and Sheela were preoccupied with arranging trays for Executive Class passengers: they did not notice the bulging pockets of the emerging passenger. And, of course, thanks to the fuss created by 60A, they had failed to notice another detail. The brown travel pouch that 59A had taken into the toilet was no longer with him when he came out.

The London-bound jumbo reached Delhi on time. The passenger who had boarded the plane at Bombay and occupied 59A deplaned, but by now the occupant of 60A was snoring in contented fashion. He knew that many a collector would happily part with a fortune to acquire that brown travel pouch now resting safe in his briefcase.

2 The Find

The sudden and unexpected ringing of the telephone shook Arul.

Who could be calling at this unearthly hour? It was 6 a.m. For the last hour and a half Arul had been busy debugging his computer programme. Despite several attempts, the monitor of his terminal kept telling him that his instructions could not be carried out. Painstakingly, he had examined the logic of his subroutines, the numerical codes for solving his equations, the data points ... all seemed correct. Yet this idiotic computer refused to accept his programme.

Yes. Idiotic! That is what a computer is, despite its advanced technology. Arul had always said so to the computer buffs in the Institute. No doubt the computer did millions of operations per second and stored billions of information bits for instant retrieval; and granted also that, but for its help, many of today's outstanding problems (including his own) would remain unsolved. But still, the computer was basically an idiot that did only what it was told to do. A fast and efficient, but unthinking assistant. That, in the last analysis, was what a computer was designed to be .... Otherwise, it would not have stopped at some trivial error of programming. It should have pointed out where the error was ... the obstinate moron. But who on earth was ringing at this hour? The Institute's telephone exchange did not start functioning before 8 a.m. and to avoid the bother of outside disturbances Arul preferred to work in the early hours of the morning. Whoever was ringing must be in the Institute, and using the internal line. Arul picked up the receiver. He was right.

'Dr Arul?' a crisp voice called.

'Yes, speaking.'

'Duty officer Shirke speaking, Sir! Trunk call for you from Bangalore … would you come down please?'

'Coming, Shirke! Hold on.' Arul rushed out.

Arul was uneasy in his mind as the lift descended to the ground floor. His father, now in his seventy-third year, was in frail health, but he insisted on living alone in his sprawling house in Malleswaram, surrounded by his rose garden. The old man would never give it up for a box-like flat on the tenth floor of a Bombay skyscraper.

Shirke, in his khaki uniform, was standing near the reception, holding the receiver of a green phone. This is where all outside calls to the Institute were directed after office hours. The other phone, blue in colour, was for internal communication only, which Shirke had used to call Arul.

'Hallo! Hallo! This is Arul … Dr Arul speaking.' He could hear the operator's voice clearly as she said, 'PP call to you sir! Hallo Bangalore … go ahead please.'

'Speak up please,' Arul shouted impatiently as he heard some indistinct tones at the other end.

'Raghavan, here, Dr Arul.' The line was now clear … and Arul was relieved to hear those words too. Raghavan! Then the call was official and had nothing to do with his father. For Raghavan was the manager of his project at Gauribidnur, where he was planning some highly sensitive experiments to test the law of gravity. In the past, too, Raghavan had telephoned from Bangalore to report on the progress of the project—but never so early in the morning. Why? He was given the answer straightaway.

'I phoned because a totally unforeseen development has taken place. We have just discovered …'

Raghavan kept on talking and Arul listened incredulously. Even the stoic Shirke could sense his rising excitement.

'Is Mr Jagdale in?'

'May I know who is calling sir?'

This counter-question always amused Arul. It clearly implied that the answer to the original question depended on the caller's identity. Why should it? But that is the way the Secretariat functions. He knew that logic and commonsense, so essential to science, need not hold sway in the corridors of power. He thus avoided treading those corridors—unless forced to do so, as on that particular day.

Raghavan's message made it imperative for Arul to leave for Gauribidnur immediately—which meant taking the afternoon flight to Bangalore. A trip to the airline counter had given him the disturbing news that, as Mister Ordinary Citizen, he was seventy-fourth on the waiting list, unless, of course, he could somehow get the VIP quota. VIPs, of course, being those with professed dedication to serving ordinary citizens like himself.

Vilas Jagdale, the Revenue Secretary and one-time classmate, was a close friend of Arul. It was to him that Arul turned in desperation for a confirmed seat from the VIP quota.

'Sir, Mr Jagdale is not in his room right now. He has gone for a meeting with C.S.'

Arul knew well from experience that this reply was one of the five stock replies. Although he knew the question to be useless, he nevertheless asked, 'And when do you expect him back?'

'Can't say! But he was to leave on a tour of Baramati with the Minister at 3 p.m.'

This was important news. He must get hold of Vilas before he took off for Baramati. Arul left his number for Vilas to call back. But he knew that, with a few exceptions, the bureaucrat does not believe in returning calls, just as he does not believe in replying to letters. So he kept calling Jagdale's number every fifteen minutes. Finally, he succeeded at one o'clock. Thanks to Jagdale's intercession, Arul was given a confirmed seat at last. To give the bureaucrat his due, Jagdale had a high regard for Arul as a scientist.

It was three-thirty in the afternoon when Arul finally packed his bag and set off for the airport. The plane was scheduled to depart at six and he was to report seventy-five minutes before departure. As he entered the terminal at four forty-five on the dot, he heard the announcement, 'We regret to announce a delay in our flight 107 to Bangalore. This flight is now estimated to depart at eight-thirty p.m., twenty-thirty hours.'

There was, of course, no trace of regret in the announcer's voice. For her this was merely a routine announcement. In any case, a monopolistic airline can afford to be callous towards passengers.

There was a time when Arul would get upset by this callousness. He had even gone as far as the General Manager on a couple of occasions to register his protests. But experience had taught him that it served no purpose, and only resulted in a waste of his time and energy. So Arul had grown not only philosphical about these delays, but had also learnt to put them to practical use. Selecting a corner chair he fished out a few papers from his briefcase and was soon engrossed in calculations.

The estimated time of departure of flight 107 as usual turned out to be optimistic. The airbus finally took off at ten o'clock and, by the time Arul entered the arrivals terminal in Bangalore, the clock was showing eleven-thirty. The faithful Raghavan was waiting outside.

Arul's suppressed excitement finally found expression as their jeep sped eastwards on the highway.

'Let's hear about your treasure chest, Raghavan. Begin at the beginning—your voice was not very distinct over the phone.'

At Arul's bidding Raghavan began in his rapid fire English, intensified all the more by the momentous news he wished to convey.

'Dr Arul, I called you because during last night's digging we encountered an unexpected obstacle … You are aware of how we are working round the clock to catch up with our schedule … last night we were down to twenty-eight metres and were planning to finish off the remaining four metres by dawn. But at about thirty metres' depth we found a layer of metal … metal so tough that our drill simply bounced off it.'

Arul whistled, but did not interrupt Raghavan's narrative which had further accelerated.

'This metallic layer turned out to be square-shaped, about three metres in size. Since the well we are digging is of a wider cross section I decided on the spot to keep digging round the obstacle … for that was possible.'

'Well done!' Arul knew that he could rely on Raghavan's initiative, which is why he had selected him to oversee this project over a host of qualified graduates.

'As we went further down, we discovered that this metallic obstacle was not a natural one. It extended uniformly down to about three metres … in fact it turned out to be a perfect cube … and appears to be made of some unknown alloy.'

'Cubical shape?' Arul had heard Raghavan mention this over the phone in the morning. But somehow the significance of it had not registered with him then. No natural rock could be exactly cubical in shape.

'Exactly a cube, Dr Arul,' emphasized Raghavan, who had obviously thought about this aspect. 'I have had it measured very accurately … it is slightly less than three metres in size. Judging from the sound it makes when tapped, it appears to be hollow. But perhaps it is a container for something valuable. You know, if this box were in Bombay, an entire family could live in it.'

'It is all very well for you to criticize Bombay's overcrowding. But remember that Bangalore, as the fastest growing city in India, is heading the same way … But seriously, Raghavan, have you opened the container, or at least had it lifted? You have the big crane still with you.'

Raghavan had suddenly gone quiet. Finally, he managed to blurt out, 'No, Arul, I have done nothing … this whole business seems to me too queer to handle on my own.'

'But why?'

'Well, to begin with, there is no lid to this box! Moreover there are strange letters and figures all over its sides.'

'Could you identify the script?'

'No! Today I spent several hours in Bangalore's libraries. I showed the script, which I had copied on my writing pad, to some experts. But apparently no one had seen it before. And in the meantime fresh trouble is brewing on the site, which is why I am glad you are here.'

'What's the problem?' Arul had a feeling that Raghavan had throughout been working his way to this point. He waited for the punch line.

'One of the technicians on the site had seen an English film about a box found during an excavation in Egypt. When the box was opened … so the story goes … a live mummy emerged.'

'Hollywood nonsense', muttered Arul. 'Surely you don't accept such fantasy?'

'I don't, but this silly ass talked about it at tea, and now no labourer is willing to come near our pit. The work is at a standstill.'

Arul gave vent to an expletive. But he still had a feeling that Raghavan had more to say, something that he was hesitating to air. To draw him out he asked, 'How do you react to this development?'

Raghavan was quiet for a while. Then he mumbled, 'I don't buy this mummy nonsense, of course. But you should see the drawings on this container.'

'What about them?'

'I wouldn't call them exactly pleasant. In fact … all of us at the site agree that they look … well … positively sinister.'

3 The Archaeologist

The lights had barely changed when the red car in the right lane raced away, making a smooth turn of ninety degrees. It was well past the intersection before the rest of the traffic recovered from this demonstration of boom and speed. Then the normal traffic lumbered along Aurobindo Marg.

'That is what I call a car … what we drive here is a bullock cart,' muttered the Sardarji in the Punjabi Hindi of Delhi taxi drivers.

'Must belong to some foreign diplomat,' suggested his passenger in envious tones.

'No sir! The car is foreign, but the owner is a Delhiwalla … all of us in Delhi know this car. It has been on the streets for the last five months.' As if inspired by the performance of the foreign car, the taxi driver stepped on the accelerator, but this hardly changed the speed of his dilapidated Ambassador.

The red Jaguar had meanwhile turned off the ring road towards Vasant Vihar. Navin Chandra Pande was justifiably proud of this acquisition of his. Ever since his school days Navin had been fascinated by speed—from racing motorcycles to fast trains, fast aeroplanes and fast cars. It had always been his ambition to own a car—not a Rolls or a Mercedes which are merely status symbols—but a really smart, fast car. During his extensive travels abroad Navin had inspected many models and studied several motoring magazines. And finally he was captivated by this bright red Jaguar. He willingly paid the heavy import duties and brought his toy home, a toy that soon became famous in the capital.

Having crossed two lanes of the smart Vasant Vihar residential area, Navin turned his car into the driveway of an elegant house and came to a halt a few metres from the garage. He took out a small box-like instrument from his pocket. It had two buttons; Navin pressed the one on the left. The garage door went up smoothly, making room for Navin to drive in.

A well equipped home in Vasant Vihar, an imported sports car, foreign trips … one would have thought that Navin was a film star or a successful industrialist. But the reality was otherwise. For despite his liking for modern conveniences, Navin was basically interested in old things—the older the better. He was an expert and highly successful consultant to the Archaeology Department. As he always pointed out, the latest devices of modern science are indispensable for research into ancient relics. The secret of Navin's achievements lay in his appreciation of the latest techniques in archaeology which he used with great flair. The numerous additions to the museums run by the Archaeological Survey bore ample testimony to his efforts.

However, even an internationally recognized expert in archaeology like Navin would have found it hard to explain how he had acquired all his wealth. Had he been married, he could have pointed to a wealthy father-in-law as the source of his material welfare. But Navin regarded himself as one of nature's bachelors, one who went as far as acquiring a bevy of girlfriends, but not a wife. So he had to explain it all as inherited income and ensure adequately that the Income Tax Department would not probe the matter too deeply.

'Well, Ram Sevak, what's new?' Navin asked his usual question as he threw himself down on his favourite couch. Ram Sevak, his trusted servant, was already setting up the decanter on the low table by the side. He knew that his master enjoyed a 'scotch on the rocks' after returning from work.

'Miss Runa called, sir. She has invited you for dinner. So I have not bothered to cook anything here.' Ram Sevak was correct in his assessment, for Runa happened to be the current favourite amongst Navin's friends.

'And, sir, a peon delivered this letter for you', Ram Sevak added, pointing to an envelope on the drinks tray.

'Fine, Ram Sevak … go and enjoy yourself for the evening.' Navin's face was benign in anticipation of his own enjoyment later in the day.

'Thank you, sir.' Ram Sevak had already telephoned a friend to get tickets for a film in a cinema house in Connaught Place.

As Ram Sevak withdrew, Navin idly reached for the decanter—when he saw the envelope. The sender's name was not on it, but a look at the monogram embossed on the back brought a frown to the benign face. Reluctantly, he opened the envelope. It contained a typed but unsigned two-line message:

> It's been a long time since we met. Must rectify the omission. See you in Sheesh Mahal, Hotel Akbar, 8 p.m.—without fail.

Without fail! Those were the operative words. The summons had come—he had to obey. Navin dialled the phone.

'Runa? … Navin here. Yes, I got your message. But … listen Runa, I just cannot make it tonight. I … Don't misunderstand Runa, it's not like that … Oh, what's the use!' he muttered to himself as he heard the abrupt click at the other end.

The Swiss cuckoo clock reminded Navin that it was seven-fifty. He got up to leave, his drink untasted.

At eight on the dot Navin entered the Sheesh Mahal restaurant. The dining room was only sparsely occupied as it was too early for the regular clientele to finish their pre-dinner drinking at the bar. It was thus easy for Navin to locate the person he had come to meet—a short, stout man in a blue safari suit.

'Welcome, Navinbhai … punctual as usual! So what will be the order of the day—drinks, dinner, or discussion? What comes first?' The man was smiling, but Navin knew what lay beneath that urbane exterior.

'Dinner, discussion—but no drinks', he replied in an even tone.

'Well, you *have* changed! But we shall see about the last part later.'

Navin quietly moved to the buffet table. He was ravenously hungry. His companion, who had to follow a strict diet, watched enviously as Navin tucked into the food. It was half an hour before Navin felt the need to talk. Looking up from a plate containing four different sweet delicacies, he turned to the business of the day.

'Pyarelalji, how is your electronics business?'

'Pyarelal was moodily stirring his black coffee. 'It is so-so … but I came to ask you how things are at your end. What is new? … Or rather, what is old? That would be more correct in your case.'

'Nothing exceptional', was Navin's non-committal reply.

'But surely, you are understating, Navin? You know how great the demand is. I need hardly remind you of the rewards … how about an AC and a stereo system for that red car of yours?'

Pyarelal knew Navin's weaknesses. In fact, he knew the weaknesses of all his contacts, which was why he had been so successful in life.

'I have to be careful, Pyarelalji', Navin replied, dealing with an eclair. 'You know how I was nearly nabbed after that incident on *Vikramaditya*. Somebody followed me from Palam to my home—I am sure of it.'

'Nonsense! You are becoming nervous without reason. These CBI fellows are absolute fools—otherwise they would have got us long ago. No, my clients abroad have long waited for some really major stuff from you. Not since those Madhya Pradesh relics ...'

'I will try. Perhaps something from the Gupta period will turn up in Bihar. But I don't think that would fetch much', Navin broke in.

'You need not worry about prices—that's my concern. Look at this list now.'

Pyarelal produced a typed list and the two were soon engrossed in deep discussion.

At about ten o'clock Navin got up. 'Well I'll see what I can do. Meanwhile, goodbye.'

'What! No drinks? Come on, let's have one to seal our agreement.'

'Not now, I am driving.' Navin turned round and moved towards the exit.

'The fellow *has* changed!' muttered Pyarelal as he, too, followed Navin.

Neither of them saw the man with a military bearing get up from a neighbouring table.

4 The News Hounds

'Raghavan, this takes the whole matter beyond our jurisdiction, damn it.' Arul was naturally peeved at this further interruption in his project. Raghavan nodded and looked at his watch.

It was three-thirty in the morning. Though it was dark, arrangements had been made to continue digging under floodlights. Round the clock work was necessary to catch up with the schedule of the gravity experiment, but tonight no work had been done. The floodlights nevertheless operated to keep the mysterious cube under scrutiny.

'Neither you nor I can claim to be an archaeologist', Arul continued, 'but even we can see that this is not from recent times. The script is totally alien. What is more, the alloy—it probably contains iron—is unknown to our technology. Look how brightly it reflects light even after heaven knows how many centuries.'

'How many, do you reckon?' Raghavan was gradually leading to an issue that he did not want to mention directly. He hoped Arul would come round to it himself.

'Can't say! But I think—no, I am pretty sure, this alloy does not belong to our post-industrial revolution times. In fact, I can safely bet that the people who made it were technologically advanced, even well beyond our level. Isn't it intriguing that the exterior of this box is so smooth that we cannot detect its lid?'

'Indeed it is! But then, these people must belong to an era well before our relics of Harappa or Egypt.' Raghavan scratched his tousled head.

'Well said! This civilization must ante-date them by several thousand years. Somehow, I imagine, all its relics were wiped out and we lost contact with it—except for this container here. Wonder what's in it.'

As Arul carefully inspected the walls of the container, Raghavan was reminded of his favourite sleuth in fiction. Sherlock Holmes would have similarly examined the surroundings of a place where a crime had been committed.

Arul suddenly burst out laughing. As Raghavan looked anxiously, he continued, 'So much for your demons! These are not of flesh and blood.' Arul was pointing to the sinister figures inscribed on the cube. This was where Raghavan had wanted to channel their line of inquiry. What were the figures?

'Of course they are mechanical monsters—robots', Arul seemed quite sure.

'But they look sinister, don't they?' Raghavan was not sure how Arul would react to this remark.

Surprisingly, Arul took it seriously. 'I agree with you, Raghavan. They do look sinister. But then, we may be influenced by our ideas of what a benign robot should look like. On the other hand, I suspect that the "artist" who drew these figures shared our reaction. Did he dislike them too, I wonder?'

Emboldened by this sympathetic response, Raghavan advanced his own conjecture.

'One normally does not associate feelings with robots—but somehow these robots don't appear to be the benevolent kind, do they?

Arul did not reply. A shiver ran through Raghavan's body as he matched his ideas with the surroundings. Of necessity, this well had been dug at a site far from human habitation. It had to provide a quiet environment for the experiment. So here they were, in a god-forsaken place, deep underground, and near a box that contained god-knows-what. If those robots decided to come out and attack them, what means did Raghavan and Arul have for retaliation or exit? He looked at the uninviting rope ladder going straight up—thirty metres of hard climb.

'No, I don't think this container has robots inside, whatever else it may have', Arul spoke out much to Raghavan's relief. 'More likely, this is a time capsule containing records of what that civilization achieved .... Well, enough of this guesswork! We will have to get this thing opened by experts and call in archaeologists to examine what is inside. This would mean contacting Delhi and inviting red tape', he cursed under his breath.

As Arul turned to climb the ladder, he discovered a plaque-like object resting against the wall. It was about a metre long and half as wide. It contained inscriptions, red in the middle, black all round. The red letters were large and few, the text in black was long and written in fine characters. The script, of course, was unknown to Arul.

'What is this? Where did you find it, Raghavan?'

Raghavan was uncomfortable as he replied. 'I doubt it has any connection with the container. In fact, we discovered it two weeks ago at a depth of five metres or so.'

'Two weeks ago?' Arul asked, surprised.

Raghavan cleared his throat. 'Obviously, we should have reported the matter earlier. By way of explanation he added, 'Well, you get all sorts of things when you dig. You were abroad when this showed up, so we merely kept it aside till you returned. Of course, I should have informed you earlier.'

'Where was it kept all this time?' Arul asked.

'Quite safe, in my office, Arul! I had it brought down here so that you could see it along with the container.'

Arul lifted the plaque, which was metallic but surprisingly light. He thought that its inscription might be similar to that on the container, but then, why was there such a difference in the depths at which they were buried? He began to climb the ladder with the plaque balanced precariously. The plaque must be kept safely, for examination later when the cube was opened.

Raghavan followed Arul with some relief. The pit where they had been was getting on his nerves. He did not look back as he climbed, but had the uncanny feeling that those monstrous robots with their sinister expressions were staring at them.

A series of discrete knocks was usually enough to wake up Arul who was generally a light sleeper. However, so deeply had he slept on this occasion that, even after getting up, he took some thirty seconds to come back to reality, only to realize that he was in a room of the Institute of Science guest house. His watch showed 10.22 a.m.

'Come in', he called out as the knocking continued.

The door opened and Vikram, the chief attendant of the guest house came in, or rather, he floated in. Vikram invariably reminded Arul of Jeeves, the superlative valet created by P.G. Wodehouse. A fan of Bertie Wooster and Jeeves, Arul always felt that Vikram manipulated his guests, even if ever so politely and discreetly.

'Sorry to wake you up, sir, in spite of your orders. But the press people would not take "no" from me. They are waiting outside.' On Vikram's face, curiosity alternated with disapproval for the intrusion by outsiders.

Arul was about to ask Vikram to seat the visitors in the lounge till he got ready, when Vikram volunteered the information, 'I have asked them to wait in the lounge, sir, telling them that you would be ready in fifteen minutes.'

Such was Vikram's way! Arul had, however, other business on his hands. He asked Vikram to book a telephone call to his director in Bombay.

'STD or demand, sir?'

'Lightning! I have the authority, you know', Arul smiled. Of course, Vikram knew. The question was a mere formality.

He wanted to consult Professor Kirtikar, the director of his institute before talking to the press. While Vikram struggled with the phone, Arul got ready and came to the lounge. He realized that he must stall till the call came through. Amongst the throng he spotted a familiar face.

'Ah, Mukund! What brings you from Bombay?' he asked Mukund, a Bombay-based correspondent of the *Express Times*. Mukund was known for picking out sensational items, be it crime, sport or political upheavals.

'Your news, of course, what else?' Mukund grinned.

'My news? What is that?' Arul tried to feign surprise.

'You need not pretend, Arul. Here, look at what the teleprinter says.' He produced a slip of paper as the others present murmured support.

'*… Dr Arul, scientist from Bombay, has discovered a relic container in his diggings at Gauribidnur …*' Arul was impressed by the speed with which news travelled.

'I sent this news to UNI last night', confessed Ganeshan, the Bangalore-based correspondent of that news agency.

'And here is our headline in the *Bangalore Chronicle*: "Time Capsule from the Past".'

Others had brought papers with similar news items. The titles were sensational, but the content thin. The journalists obviously wanted more to follow.

'May I first know how you learnt of this event?' Arul felt it was safer to ask questions himself—the surest way of stalling news hounds.

'Well, a local correspondent of our paper happened to be in Gauribidnur, where he heard one of the labourers on your site holding forth in a tea shop.'

'You are sure it was a tea shop and not a beer bar?' Arul quipped. There was general laughter. But Kumaraswamy of the *Chronicle* was persistent.

'Jokes apart, sir, may I add that I myself went to your site and talked to Raghavan, your manager? Apparently he has sealed access to the pit.'

'What did Raghavan tell you?' Arul asked.

'He refused to comment.'

Reliable Raghavan! Arul had to emulate his discretion. Questions now started pouring in. Why this secrecy? Why was Arul here? Where was he last night, since he had been sleeping something off? ...

'Sir, call for you from Bombay.' Vikram had materialized exactly when needed.

Arul turned to the gathering. 'Gentlemen! I cannot comment till I talk to Bombay. And then, too, I can give out only what I feel should be disclosed .... Please remain seated, and I shall ask Vikram to provide you with tea, coffee and biscuits.'

'It is all ready, sir!' Vikram quietly wheeled in a laden tea trolley.

5 The Task Force

'The D.G. wishes to see you', said the note on a slip of paper marked UR-GENT. The signature was illegible but familiar to Navin from past experience. He contacted the P.A. to the Director General on his internal phone.

'The D.G. is presently busy talking to Bombay long distance; but please go in right away, sir.'

The D.G.'s office was at the end of a long verandah. Navin paused briefly as he passed through the P.A.'s room.

'Shankar, any idea why the summons?' In the past Shankar's briefing had often been useful.

'Can't say Mr Navin. But he called you soon after a call came from Shastri Bhavan.'

Shastri Bhavan—the seat of the Ministry's secretariat. So the matter was 'official'.

'And then He asked to speak to Professor Kirtikar in Bombay.' Shankar's reverence for his boss required 'he' to be pronounced with sufficient gravity to justify the capital 'H'.

'Who is this professor?'—the name was unfamiliar to Navin.

'The director of the Basic Research Institute in Bombay, Mr Navin. He is presently on the line.'

It suddenly dawned on Navin that the 'urgent' matter must concern the news item that had interested him in the morning paper. Wasn't the discovery of the big container made accidentally in the course of excavation for some experiment of this institute? He entered the sanctum sanctorum after a light knock on the teak door.

'Yes, Professor Kirtikar! We will move right away—please keep everything intact until our experts come … bye!'

The D.G. smiled at Navin as he placed the receiver in its place. His Buddha-like face was grave as well as benign.

'Have a seat, Navin … You probably know what it is all about … that container at Gauribidnur. Last night both PTI and UNI wanted to know what it was all about. Naturally, I could not comment, as I did not have any first-hand report myself. But I contacted Professor Kirtikar, the director of the institute in Bombay. All he could tell me was that their scientist in charge, Dr Arul, was on his way to Gauribidnur and that he was awaiting his report.'

The D.G.'s face now wore an amused smile. Navin wondered why. The elucidation was not far off.

'But last night's experience warned me that events were going to develop further', continued the D.G., trying to keep a straight face. 'So, to anticipate future action, I immediately called Harisharanji at his home.'

Harisharan was Secretary for Culture in the government. Navin was familiar with this official and could now guess why the D.G. was amused. He lighted a cigar and listened for further details.

'I asked Harisharanji for permission to send an expert task force to Gauribidnur right away. But he was cautious. He pointed out that we are already well into January and that until the new financial year starts it would be injudicious to undertake any new expenditure.'

'Rubbish! This rule goes by the board when it concerns his own foreign tours.' Navin could not control himself.

'I know—but listen to the end of the tale. I got a long lecture from him on Plan and non-Plan expenditure, the cuts imposed by the Planning Commission, economies to be achieved, and so on. I began to feel as if I had stirred a hornets' nest.'

Navin was not surprised. Mr Harisharan was, in his (somewhat biased) opinion the typical example of a senior government officer—ultracautious to the extend of doing nothing new, always sheltering behind procedures, and bound by red tape. Exceptions exist, Navin was the first to admit. Indeed, he knew of some enlightened secretaries who were efficient and saw beyond the maze of rules and regulations. But these, in his opinion, were a handful. He recalled his visits to the Lal Bahadur Shastri Academy in the pleasant surroundings of Mussoorie where he had spoken to civil service probationers. He had found them an enthusiastic lot—full of intelligence, initiative and freshness—just what the services required. But what the service moulded them into was the stereotype that was Harisharan. Why should the Archaeological

Survey be burdened with a Harisharan rather than have the enlightened guidance of a Probir Ganguly? Probir Ganguly was one of the handful of exceptions—but he had been moved on to the Home Department after a brief stint at Culture. Now they were saddled with Harisharan!

'And then this morning the situation changed dramatically', the D.G. added with a mischievous smile. Harisharanji received a phone call from the South Block. No less a person than the P.M. was on the line!'

Navin whistled. He could imagine the scene ...

'The P.M., it appears, had read the news in the morning papers. He was quick to spot that the finding of the container was of great significance. He asked Harisharanji what his department was planning to do about it.'

Navin let out a guffaw. 'I bet it woke up Shastri Bhavan.'

'It did! In a complete volte-face, Harisharanji rang me up a little while ago. Now, it is all systems go! We have to send a task force to Gauribidnur right away ... and that's why you are here.'

The D.G. banished the smile he had picked up while narrating the incident and sat down. Navin knew that the ball was now in his court. He flicked the ashes in an ashtray and said: 'Frankly, my first impression was that all this was a big practical joke. But on second thoughts I reasoned otherwise. What practical joker can place a three metre cube container a hundred metres underground? It's impossible! But then, the other conclusion, that we have unearthed a relic of some ancient civilization, is also improbable. However, as scientific investigators we should not jump to conclusions. I will leave for Bangalore tomorrow.'

'Today, not tomorrow! Go by the afternoon flight via Hyderabad. Take whomsoever you want with you. And ...'

The D.G. paused, knowing that what he had to add would not go down well.

Navin guessed as much as he waited for the rest.

'... In Hyderabad, Dr Laxmanan will join you. I will send him a telex right now and also brief him on the phone.'

'Laxman? Why do we need Laxman in this business?'

Navin, as expected, was irritated. This was to be his show—his entirely. Why should he have to rope in an outsider?

'I feel pretty strongly' said the D.G., emphasizing the last word, 'that we may sooner or later need someone who works on codes, languages ... and even artificial intelligence. No one but Laxman can handle it.' He pressed a bell to summon Shankar.

Navin moved to the window overlooking one of the tree-lined avenues for which New Delhi is so famous. He heard Shankar enter, take down the telex message and depart.

'Where archaeology is involved, you are in charge of the whole project, naturally', the D.G. added in conciliatory tones. 'Dr Laxmanan will be joint project leader along with Dr Arul, who is already there. Professor Kirtikar, to whom I was talking before you came, has agreed.'

Arul was an unknown quantity in Navin's personal equations. But Laxman was a difficult customer. Navin looked upon him as an untamed horse. It was hard to work with a man who was brilliant and at the same time totally independent in approach. Laxman always preferred to go his own way.

And what use would an AI expert and a code-breaker be in such a case? Codes can be deciphered using some knowledge of the thinking and culture of the sender. Here—if the container did indeed turn out to be genuine—they were dealing with a civilization far removed from the present one. So far as artificial intelligence was concerned, Navin knew little about it, but he could see no relevance of it to the present project. But he had to humour his D.G. Then a thought crossed Navin's mind and put him at ease. A man of Laxman's restless brilliance would surely be bored and quit if he found that he was not relevant to the project!

'Sir, I foresee a few practical problems right away', he turned to the D.G. 'We will have to take the container out and examine it under tight security. Where can this be done?'

'I discussed this point with Kirtikar. It seems they have just completed a building for Dr Arul's project on the site. It was meant for labs and a computer, but it is still unoccupied. The container can be housed in a hall earmarked for one of the workshops', the D.G. clarified.

'And the computer? Laxman cannot work without one, you know.'

'A VAX model has already arrived and will be installed within three weeks once the air-conditioning is straightened out. And so far as security is concerned, since South Block is interested, we will get all we need. In fact, I will get in touch with the Home Ministry immediately.'

Blast! This was not what Navin had meant. Would this bring the CBI to his doorstep? Anyway, that could not be avoided now.

They discussed further details for an hour during which the D.G. contacted Probir Ganguly, Harisharan and others. When Navin entered his office, his steno, Rajan, had a message for him.

'Someone who did not leave a name asked you to call this number.' Rajan did not notice the frown that crossed Navin's face as he glanced at the number.

Although he could dial outside calls from his office phone, Navin did not do so. He left his office at lunch, leaving a message that he would fly to Bangalore in the afternoon.

But he did not go for lunch immediately, first going to a public telephone booth and dialling the number contained in the message.

'Pyarelalji, I told you never to ring me at my office', he began aggressively.

'Sorry, Navinbhai!' Pyarelal spoke in his soft voice with mock regret. 'But the matter was urgent and you had left home.'

'Well?' Navin was afraid, and knew what was coming. Nobody was better informed than Pyarelal.

'I gather that the P.M.'s office has asked your department to investigate this container at Gauribidnur.'

Navin was silent. It was never feasible denying what Pyarelal's spies had told him. Pyarelal continued:

'Well it is just a gentle reminder … don't forget me when you are out there.' The phone clicked off.

The conversation made Navin tense. Why were things becoming so complicated?

There were seven clocks in Pyarelal's den, each showing a different time. The names Tokyo, Singapore, Bombay, Dubai, London, New York and Los Angeles identified them. It was on the fifth one that Pyarelal's attention was focussed. It showed the time at 8 a.m.

He turned the dial of his phone to get a London number. He had greatly welcomed the introduction of international subscriber dialling, for it left no written record with any intermediary operator. With five attempts he finally got the number, and after several rings someone at the other end replied sleepily, 'Who the hell is it?'

'P.L.' replied Pyarelal in subdued manner.

'Hope you have something really important to say to justify waking me at this hour' said 'London' in a quieter but still menacing tone.

Pyarelal expected this response, but was confident as he replied:

'Not to be relayed on the phone. But look out for news from India on Breakfast TV—and take appropriate steps. Call me after ten hours.'

Pyarelal hung up. At the other end 'London' sat up in bed and switched on BBC1 by remote control. After an interview with a Ford executive and the latest report on the lock-out at the motor company headquarters, the programme turned to a chat with a cabinet minister—the one for trade and industry. On ITV1 a dog trainer was describing her methods of teaching various tricks to dogs. 'London' switched back to BBC1 at 8.30 a.m. After the Ford strike, the run on sterling and opinion polls in a marginal constituency, the newsreader finally came to what 'London' was awaiting:

'In southern India, in the small town of Gauribidnur near Bangalore, archaeologists are examining an excavated container believed to be several thousand years old. No official comment has yet come.'

The news turned to football prospects. 'London' promptly switched off the TV set and jumped out of bed. In the next ten hours he had to get a lot done.

6 The Expert

Laxman read the telex once again and reluctantly reached out for his phone. Reluctantly, because he was afraid to face the reaction at the other end. He dialled.

'Umi … Laxman here. Look, something has turned up.' He came straight to the point. The expected reaction also came.

'Whatever it is, you have to come on time … we are going out at eight sharp.'

Laxman grew more uneasy. He knew how shattered Urmila would be by his upsetting the programme—their first wedding anniversary. But he could do little to calm her and decided to get it over with.

'Umi … you have to forgive me, but that is off. I have to go to Bangalore by the evening flight.'

'Oh no!' he could almost sense the imminent breakdown. He spoke rapidly. 'You read this morning about the container found at Gauribidnur, didn't you? Well I have to rush over there to investigate.'

'But what have you got to do with it? You are no archaeologist.' Urmila's tone was a mixture of anger and despair. Laxman persisted. 'No, this case is very unusual. They may need my skills to decipher the contents of this box … Look, Umi, shall we leave it open tonight? … I will settle it with interest when you join me there.'

'Join you? So it is going to be a long drawn out affair! And what am I to do in that god-forsaken place all by myself, while you are busy with that wretched container?' Sniffles at the other end warned Laxman to be ready for the explosion. He produced what could be a trump card: 'Listen, Umi, I am on my way home now. On the way I will pick up that sari you set your eyes on—OK?'

Urmila's response was distinctly modified. She had wanted that expensive silk sari but had not expected it even on her wedding anniversary.

'Trying to bribe me, aren't you? Well, we will discuss everything then. So you will come for lunch—that will be a change! Come soon.'

As she hung up, Laxman knew that he was forgiven. He began to put the finishing touches to his work before leaving his office. That Dr Gupta, the Director General of Archaeology, should have asked him to join the task force reflected on his keen perception of Laxman's ability. A PhD from MIT in computer software development, Laxman had had a bright post-doctoral career at Caltech. His research papers had made him internationally known and he could very well have joined one of the multinationals. But he always felt that his roots were in India and he had come back to take up a post specially created for him by Hind Electronics at Hyderabad. He was allowed full freedom to do his research and to go anywhere at any time if it suited him, and his contributions to R and D had amply vindicated this unusually liberal policy. Even his rivals in the field acknowledged his superiority.

But Dr Gupta's request to go post-haste to Gaurbibidnur convinced Laxman that there was more behind the short news item he had read in the morning. If this box turned out to contain technologically advanced relics of the past, one might have to modify the accepted history of human progress. He had read and dismissed as speculative a highly readable paperback which purported to show that an extremely sophisticated prehistoric civilization had once existed on Earth. If this container was genuine, it might contain evidence that could not be ignored. Laxman eagerly looked forward to his new assignment.

The telex informed him that his ticket could be collected from the ariline's office at the airport. So he had about forty-five minutes to pick up the sari on his way home to lunch. He closed his briefcase and made his way to his old and decrepit car in the car park.

7 The Agent

The British Airways jumbo from London landed at Sahar at 3.20 a.m., precisely at its scheduled time of arrival. Most of the passengers were bound for the Far East and Australia. Of the few who got off at Bombay, a hefty white man seemed to know the airport well enough to make his way quickly out of the hold-ups at immigration and customs. Even so, by the time Karl Shulz came out of the arrivals hall it was nearing quarter to five. Making his way through the melee of passengers from the Gulf states and their enormous cases, Shulz walked along the long pavement by the terminal building. He had only a small hold-all; he never believed in travelling with checked-in baggage. The crowd thinned appreciably as he walked a hundred metres, avoiding or turning away unauthorized taxi drivers and hotel touts. His brisk walk turned to a saunter and finally he came to a standstill almost at the end of the pavement. He was whistling the signature tune of 'Dallas'.

'I heard there is a new actress playing Miss Ellie now … I am Mahesh Doshi. Call me Mahesh.'

The speaker was an Indian of average height and build, looking almost diminutive beside Shulz. Shulz stopped whistling, handed his hold-all to Doshi and followed him to a red Maruti standing in the car park.

'Call me Karl', he said barely managing to squeeze into the front seat.

'Karl, it is now nearing five. Your flight for Bangalore leaves at seven. Shall we have a drink at the Centaur?' Mahesh glanced at his watch as the little red car sped along the road to Santa Cruz.

'Let's go to the coffee shop.'

The coffee shop, open for twenty-four hours, was nearly empty … apart from two sleepy Russians and an Indian family; the latter were obviously from the USA, judging from the complaints of the two children speaking in American accents.

'I told you not to go by Air India … it is always late!' The mother had evidently not forgotten her native Marathi. 'Come on Neela, you exaggerate. We had no better experience with Pan Am last time. And our travel agent offered the best deal on Air India … Remember we saved four hundred dollars, more than five thousand rupees.' The husband justified their choice of airlines.

Mahesh smiled … it was characteristic of NRIs to convert to rupees, even after several years abroad. The children, however, were rooted in America.

'Maa…mmy, I want coke! I don't like this substitute', the older one complained.

'When canna have Kentucky Fried Chicken, Mommy?' the younger one, a five-year-old, asked. He had already decided that he had had enough of the Indian delicacies showered on him by grandparents, aunts, and uncles.

'Generation gap!' muttered Shulz and then turned to Doshi: 'I bet you are waiting for your green card.'

Mahesh smiled with some embarrassment. Shulz had guessed correctly. He was waiting to follow his two brothers and sister to the land of hope and glory.

'Coffee for me; black, without sugar', Shulz ordered. Mahesh asked for a cold drink. They had selected a secluded corner.

'My names may change, but my taste for black coffee remains the same', Shulz smiled as he recalled the many aliases he had used.

'To come to business, Karl: we have fixed you up in the Royal Manor, as you wanted', Mahesh began. 'You can rest in the afternoon to sleep off the jet lag. But do you really want to pay a visit to Gauribidnur tonight?'

'Of course, I must. What's new there that I must know?'

Mahesh was dreading this question. He hesitated before replying.

'Nothing to report, I'm afraid. We just can't penetrate the tight security. Now they have got the C.S.F. patrolling the site, inside and out.'

'C.S.F.?'

'The Central Security Force! And besides, there is strict censorship on all news from the Science Centre. Even the labourers who found the box have been replaced … they have been taken to god knows where!'

'And Raghavan?'

'He is still there, very much in charge. But totally beyond reach … But if I may hazard a guess, they will soon open the container.'

'I want facts, not guesses.' There was only a passing shade of displeasure across Shulz's face, but it would have been enough to give some indication of the real person that lay beneath that suave exterior. Mahesh missed it as he sipped his cola.

'My guess is based on facts … supplied by Navin a couple of days back.' Mahesh felt himself on firmer ground here. Navin was the chink in this armour around the container, their only hope.

'Aach! Navin, of course! Haven't seen him since we flew together from Bombay to Delhi. Must renew our friendship', Shulz muttered softly. 'Mahesh, please arrange for me to meet Navin tonight.'

Although prefaced by 'please', Mahesh realized that it was an order. He felt out of his depth.

'I will do my best', he said uncomfortably.

'That's not good enough, my friend! You have to ensure that it comes about … come let's go.'

As Shulz's heavy hand fell on his back, Mahesh shuddered. It would not do to make this man angry, he realized. He gulped down the drink and got up.

Karl Shulz reached Bangalore on time. Making his way through the mob at the exit of the arrivals lounge, he found what he was looking for—a man holding up a small board with his name on it. He followed the man to a black Ambassador in the car park. Its black number on a white plate told him that it was hired from one of the private taxi firms in the city.

'Royal Manor, please', said Shulz, although he knew that the man had been properly briefed beforehand.

'Yes, Saar! Pyarelal Saab wait you, hotel … lunch.' The driver managed to convey the information across the language barrier.

Skirting the golf course the car made its way to the imposing pseudo-British building. A spacious suite was reserved for Shulz. He lost no time in refreshing himself with a long, cold shower. When he came out of the bathroom, he put on a kurta-pyjama made to measure. The outline of his well-built body was noticeable through the semi-transparent Lucknavi kurta.

The bedside phone rang.

'Shulz', he acknowledged, speaking into the receiver.

'Welcome, Karl!' Pyarelal here. Hope you got my message for lunch. Is it OK?'

'Where?' Shulz did not believe in talking too much on the phone.

Pyarelal named a leading Bangalore restaurant. 'Be near the reception. A car will come to pick you up.'

'When?'

'Twelve noon … OK? … Bye!'

There were no outward signs of fatigue on Shulz's face as he sat opposite Pyarelal, devouring Tandoori chicken. It was characteristic of the man that he could will himself to be in good physical condition whenever the occasion demanded. His working phases of iron self-discipline alternated with those of relaxation when he let himself go, which is why he declined alcohol, and came straight to the point.

'I am upset by what Mahesh told me.'

'What did he tell you?'

'That you are unable to penetrate the veil of security around the container.'

Pyarelal's face had an enigmatic smile. 'Karl, in these matters it won't do to rush things … I have hopes in Navin.'

'Hopes! What can Navin do?'

'Navin is silent because he is awaiting developments. Believe me, he will find a way of communicating once he has something to report.'

'Or is he having pangs of conscience?'

'He has had them before', admitted Pyarelal, 'but I cured him of those.'

'Well, I was planning to meet him tonight.'

Pyarelal's face fell as he heard this announcement. 'Karl, I need you to be incognito, at least as far as the Science Centre lot are concerned', he added. 'In Bangalore you would pass off as a tourist—but in a small place like Gauribidnur …'

'I will stick out like a sore thumb, won't I?' laughed Shulz.

'And, besides, you might lose your temper. Navin needs to be handled delicately, especially if he is passing through a mental conflict.'

'Anyway, I will visit the site tonight, for sure. In deference to you, I will not see Navin—but I'm sure you cannot object to a nocturnal visit by me?'

'Who am I to object? I will arrange for you to be collected tonight at ten. Then it is up to you.'

Shulz glanced at his watch, which still registered 8.30 a.m. GMT. He made a mental calculation—it was now two in the afternoon. He had about seven hours of sleep ahead of him. … Fortunately, he could sleep at will anywhere, any time. His job demanded it.

He nodded and got up.

8 The Opening

'What do you think, Laxman?'

Arul's question came nearly an hour after the discussion had started, an hour during which Laxman had hardly spoken. Take any committee and you will find its discussion dominated by two or three members, and it is not usual for such people to have something relevant to say. Today's discussion was essentially a monologue by Navin, which frequently took off in a tangential direction. Arul, who vainly struggled to control the situation, finally had to appeal to Laxman, knowing that he rarely saw eye to eye with Navin.

'I agree with Navin.'

Laxman's reply surprised everyone, not the least Navin. To elaborate further, Laxman continued, 'We have spent enough time examining the container from outside. We are not the least bit wiser. Now we are wasting time discussing the possible dangers of opening it. The scientist in me says "Open it—and to hell with the consequences". Those who left the container behind clearly intended the finder to open it and examine its contents. Otherwise, what was the purpose in leaving it at all? If you people are afraid to be around at the time of opening, leave the job to me.'

'Well spoken mate', Navin added as he rose and shook Laxman's hands. 'I for one volunteer to be with you when it is opened.'

'I too', said Arul.

'But how are we going to open it? We have not yet found even traces of a lid.' The question, of course, came from none else than the practical Raghavan.

All the attempts so far to discover the container's lid had indeed been unsuccessful. The container had been taken out of the deep well where it was found and placed same side up in the large hall of the workshop. All manner of detecting instruments, including electronic devices, had failed to reveal any discontinuity on its smooth surface. Brute force, in the sense of explosives, was ruled out as it might damage whatever was inside.

The committee's discussion thus turned to the practical issue of opening the container and it was Arul's turn to remain intriguingly silent. But an idea had occurred to him. Indeed, so quiet was he that nobody noticed when he left the conference room. Thus, it came as a surprise to all present when he made a dramatic entry, chanting 'Eureka'.

'Gentlemen! I think I have the solution to Raghavan's problem.' He smiled at the profound effect this sentence produced on the group. He continued:

'We could not open the cube because we were ignoring the most obvious clue.'

'Nonsense! We have examined every inch of that wretched surface as minutely as we could.' Laxman's rebuttal found sympathetic echoes round the room. Arul experienced an impish delight in stringing them along.

'No doubt you have read Edgar Alan Poe.' He spoke quietly.

'Speak to the point, Arul. Who is this Poe?'

'Navin, didn't you ever read mystery stories, the whodunnits?'

'Of course! Right from Sherlock Holmes to modern-day thrillers—I can claim to be reasonably well informed', Navin rejoined.

'But you did not go as far back as Poe, who could be said to have started this kind of literature in modern times ... Let me tell you of his story of the purloined letter. A compromising letter had to be retrieved. But the person who possessed it had hidden it very cleverly. The searcher ransacked the entire room looking for secret compartments and drawers—but the letter was not to be found, so cunningly was it hidden.'

'Where was it?' asked Laxman.

'Exactly where nobody would think of looking! It was lying open on the desk.'

'I see what you are getting at. We have missed an obvious clue ... but what?' asked Navin, exasperated.

Arul laughed. 'Let me demonstrate it to you! Come to the container.'

Everyone trooped along to the workshop. Raghavan opened the doors with his special key. The container lay there as if defying the collective intellect of the twentieth century.

'Look at these pictures outside. What do you make of them?' asked Arul. 'They are not purely decorative but have a function to serve. Could they not be intended as instructions for the finders of the cube, who would not be expected to know the language of its creators? ... This is what I have been trying to work out ... especially the pictures on the top.'

Arul led the way to the overhanging gallery erected to provide a view of the container from the top. He addressed a general question:

'Take that square being pulled by two elephants on opposite sides. What does it convey to you?'

'Perhaps those people used elephants to drag heavy weights', Raghavan hazarded a guess.

Arul laughed. 'If so, why should the elephants be shown pulling the block from opposite sides? And do you expect such an advanced civilization to employ animals for mechanical jobs?'

There was silence. So Arul continued. 'This square in the picture is, of course, our container. That two elephants on opposite sides are unable to pull it apart is what we are meant to deduce. Our own history had a similar episode except that there were horses instead of elephants.'

Laxman was the first to catch on. 'Of course ... the two hemispheres of Magdeburg.'

Several amongst those present were still blank, so Arul elaborated. 'Back in the seventeenth century, the scientist Otto Von Guericke joined two hemispheres, evacuated the space in between and then tried to pull them apart with horses. The air pressure on the hemispheres was so great that the two hemispheres just could not be separated.'

'Which implies in the present instance that the box is vacuum sealed', Raghavan added. 'But then, how was the air taken out from within?'

'Obviously through a small hole which must be concealed somewhere. The pictures again should provide a clue. May I draw your attention to these ellipses?'

Several ellipses with different shapes and orientations were superposed at one corner of the top. There seemed to be no symmetry in the picture. As they tried various conjectures someone remarked: 'It reminds me of the trajectories of comets in our solar system.'

The remark did not inspire the speaker to any interpretation. But it was Laxman again who burst out, excited:

'Isn't it obvious? Like the Sun, which is the common focus of all cometary paths, all these ellipses must have the same focus.'

'Exactly', said Arul. 'While you were arguing I slipped out to look at our photographs of these drawings. Here is one of the ellipses.' He produced an enlarged photograph, on which he had done some geometrical constructions.

Every ellipse has two focal points. The ellipses drawn in Arul's diagram had the remarkable property of all sharing one common focal point.

'The makers of this box did not wish it opened by primitives. They expected some mathematical knowledge on the openers' part. Thanks to Arul, we qualify.' Navin's reasoning appealed to everyone present.

Raghavan procured precise measuring instruments to determine the exact common focal point on the box. This was the point where the evacuation would have been done. However, electrical drills failed to puncture the surface. Arul himself was nonplussed. Was his inspired reasoning to prove a red herring after all?

'I think the drills we are using are too thick … I will try a very thin needle', Raghavan spoke as if suddenly inspired.

'Well done, Raghavan!' Arul slapped him on the back and sent him to fetch the finest needle available.

And to everybody's delight the finest of all needles pierced the metal as easily as it would a piece of cork. There was the hissing sound of air entering the container as the needle was withdrawn. The entire top of the container automatically came up by two inches and turned on a hinge.

The contents were covered by a fine cambric-like material. As it was pulled aside, everyone present had the feeling that he was looking at something out of this world.

9 The Committee

A letter in an envelope marked 'secret' and closed with sealing wax, contained within another sealed manila cover, and that too delivered by special courier … Professor Kirtikar as a rule was not a frequent receiver of such mail. He read the contents with some misgivings for he never looked forward to committee meetings in the nation's capital. The letterhead carried the address of the Department of Science and Technology in red letters to indicate that it was an official communication from a department of the government.

> A top level committee has been constituted to look into the findings of the Gauribidnur container and you have been appointed a non-official member of this committee. The committee's first meeting has been arranged in Technology Bhavan at 11 a.m. on 2 February and you are requested to kindly make it convenient to attend the meeting.
> An office order specifying the TA/DA rules for non-official members attending the meetings of this committee is attached for your information.

Professor Kirtikar smiled as he read through the letter. The contents were all phrased in the passive mode so cherished by bureaucrats. The addressee must never know who was to be held responsible for all the plans reported in the letter—certainly the sender could not be held accountable for the statements in it. The only thing that could not be avoided by the sender was providing his own name (albeit slightly obscured by an illegible signature) at the end of the letter.

Raj Nath! Kirtikar mused sadly as he read that name. Raj Nath, a one-time colleague on the institute's faculty, was a close friend of his. And of so many others, young and old. A lively person with liberal views, Raj Nath had been affectionately called 'Smoke Chimney' by his colleagures, for his habit of incessantly smoking through a pipe. Indeed, it was often difficult to see Raj Nath clearly through the smoke screen around him. But those who had discussed science with him knew that beneath all that smoke there was a highly perceptive brain. A molecular biologist by profession, Raj Nath had views on fundamental physics ranging from superconductivity to cosmology and would often be found animatedly expressing them in the institute's canteen.

But, alas, not any more! Just over two years ago, a short while before Kirtikar himself became the director, the long arm of the government, always on the look-out for distinguished scientists to run the science administration from New Delhi, had taken Raj Nath away to head the D.S.T. His colleagues were sorry to see him go, but had hoped that a man of his freshness and liberal outlook was just what the bureaucracy in Delhi needed.

As Kirtikar looked at Raj Nath's letter he realized how misplaced those hopes were. The same Raj who used to be infuriated by officialdom, now excelled in writing DO's in the best officialese. Here was yet another promising scientist eaten up by the bureaucratic Black Hole of New Delhi, thought Kirtikar as he glanced at the names of the committee members.

The high powered nature of the committee was obvious from the fact that no less than the Home Minister, Bhagvati Dayal Upadhyay, a minister of cabinet rank, was chairing it. The Minister of State for Science was the next person, followed by secretaries from the Departments of Home Affairs (Probir Ganguly), Culture (Harisharan), Information and Broadcasting (Shafi Ahmed) and Science and Technology, represented by Raj Nath as the Convener. The list of official members also included Dr Ramesh Gupta, Director General of the Archaeological Survey and, curiously enough, two other names with no designation given. Of these two, the name of Major Samant seemed to ring a bell, but exactly when and where Kirtikar could not recall.

The unofficial members included, apart from himself, Drs Arul, Laxmanan and Navin Chandra Pande. A Raj Nath touch, probably. Otherwise such junior people would never have found their way into a high-level committee of this kind. Well ... there is still some fire left in the old dog after all, thought Kirtikar, as he called for his travel section to book him a ticket to Delhi.

The date of the meeting, he noticed, was the very next day.

Technology Bhavan is a single storeyed building standing next to the Qutub Hotel on the outskirts of Delhi. Unlike most other government departments which are in the neighbourhood of Rashtrapati Bhavan, the Science and Technology Department was tucked miles away from the corridors of power. Was this symbolic, Kirtikar used to wonder?

The building itself once belonged to the United States Information Service. Now maintained by the P.W.D. it had naturally lost the polish it had in earlier times. Kirtikar, who had visited the U.S.I.S. a couple of times in the past, could not fail to notice the decline in standards as he was conducted to the Secretary's office.

'Hallo, Raj!' he greeted the figure barely visible behind the smoke. He was meeting Raj for the first time since he had left for this Delhi assignment. He was somewhat taken aback to see the change in his appearance.

'Welcome Prashant ... take a seat', Raj Nath greeted him with the characteristic effusiveness that came so naturally to him.

'You startled me, Raj. When did you grow such long hair? You look like one of those ancient sages living out in the jungle', Kirtikar said, half jokingly. But Raj turned serious.

'Whether I look like a sage or not is debatable. That New Delhi is a jungle, is not! ... Come, tell us about good old Bombay, which I miss so much. But, first, tea or coffee?' He pressed a bell fixed to the side of his table.

His P.A. entered, duly took the order and departed. Yes! Things were different here. In earlier times they would both have trooped down to the canteen, stood in the queue and served themselves.

'Good you came somewhat earlier, so I can brief you about this meeting', Raj Nath relit his pipe.

'You had better! In any case, a rustic from Bombay like me feels a little overawed by this high level committee.'

'It was constituted by the P.M. himself ... in fact, it went through an amusing metamorphosis. Strictly between us, I will tell you how.'

'Absolutely!' Kirtikar recalled how often he had heard that phrase from Raj back in the institute. He knew it to be a prelude to some scandal.

'The P.M. wanted an expert committee to quickly assess the container and its contents; and he so instructed the Department of Culture. Naturally, the matter landed in Harisharan's lap.'

'Harisharan?' Kirtikar asked.

'Secretary, Department of Culture', explained Raj Nath, blowing out a smoke ring. 'Harisharan promptly constituted a list and sent it to the P.M. for consideration ... you know what the P.M.'s comment was? He said it looked like a marriage party made up of caterers and bandsmen but without the groom and bride.'

How did Raj Nath know what the P.M. had said, wondered Kirtikar. Was this part of the grapevine for which Delhi was so notorious?

'Apparently Harisharan's list included secretaries and additional secretaries from four departments, state officials from Karnataka and the Collector of the region including Gauribidnur.' Raj Nath continued with a smile, 'In short there were no real experts on the committee ... And so I was summoned to South Block and asked to constitute this committee. You know the result. Harisharan would have had a fit to see such juniors as Arul or Laxman on the committee; but he had to go by the P.M.'s decision.'

Raj Nath then started briefing Kirtikar about the other committee members from Delhi so that he would be on guard. As he was halfway down the list, he looked at his watch and rang the bell once again.

'*Are Bhai, chay ka kya hua?* The time for the meeting is drawing close', he told his P.A.

'It's coming, sir! I will telephone the canteen again, sir.'

'Well, that's Delhi for you, Prashant', said Raj Nath as the P.A. left. 'It's all bound up with who is at what level and who can do what. I cannot go to the canteen myself—the whole deparatment would be shocked if I did. So I tell the P.A. It would be below his dignity to go there now. So he will send a peon ... and so it goes on. You notice the contrast in efficiency even more when you come from our institute.

That's the tragedy, Kirtikar thought. Efficient people from well run places are called here just to be eaten up by the 'system'.

The tea came soon, however. But they were destined not to finish it. For halfway through, word reached Raj Nath that the two ministers were due any minute. He left his cup and rushed to the front gate to receive the V.I.P.'s, while Kirtikar ambled along to the committee room.

The meeting started on schedule with the chairman calling upon members to introduce themselves. It was then that Kirtikar learned who Major Samant was. After Samant had introduced himself as 'coming from the Intelligence Bureau', the chairman felt the need to elaborate further.

'Major Samant has been characteristically reticent. Perhaps I should add that he has been awarded the Veer Chakra for his bravery in the Bangladesh War of Independence. He has since then done a lot for the I.B. but, of necessity, the details cannot be disclosed. I am happy that he is in charge of security at the Science Centre.'

All but one in the committee were reassured by this description, for they shared the concern for security and confidentiality about the newly found container. All but Navin, that is. Navin had already experienced the effect of Samant's efficiency. Just how much did Major Samant know about his past?

An incident came to Navin's mind, something that had happened a couple of days back. He was getting ready to go to Bangalore for a dinner engagement when Major Samant jokingly remarked, 'Out for a date, Mr Navin? I am sure you have found a loved one in Bangalore.' This was spoken in Hindi with the word *pyari* used for 'loved one', and Navin had laughed it off. But now he began to wonder—why had Samant used that particular word? Did he know about the Pyarelal connection?

A pat on the back brought him back to the present. The chairman was addressing him.

'I beg your pardon, Minister ... I was not paying attention,' he apologized.

'Mr Pande, we have been told that the container was opened last week. Can you give us an account of how this feat was accomplished? The minister repeated his question.

As head of the archaeological project, it was Navin who had to present the account. This he did in his fine narrative style. For technical details he gave way to Laxman and Arul. By way of conclusion, he added, 'We can now open and close the container. But we are awaiting this committee's sanction to proceed further and investigate its contents.'

'I am surprised, Mr Navin! With your archaeological expertise and with a team of scientists to help you, how could you contain your curiosity?' the minister asked. 'I would have expected you people to have ransacked the container right away.'

Navin, Arul and Laxman glanced at one another, as if avoiding the answer. It was left to Kirtikar to come to their rescue.

'Sir, they would indeed have done so but for a restraining order from Delhi', he clarified.

'Restraining order? Why? Who sent it? The chairman looked round the table.

'Er … I gave the order, Minister', Harisharan mumbled, somewhat embarrassed. 'Realizing that it was a sensitive matter I felt that all investigations should proceed through proper channels after due clearance from this committee. This is standard practice, sir. I simply followed it.'

The minister smiled. If Harisharan's house catches fire, what and whose clearance would he require before calling the fire brigade, he wondered? Aloud, he said 'I can understand your approach Harisharanji, as one dictated by caution … But surely all here would agree that these experts at the Science Centre are mature enough to decide for themselves? They don't need to run to this committee for every small step.'

Everybody agreed.

'I will so minute it Mr Chairman!' added Raj Nath, tongue in cheek.

There was, however, a discussion on what should be done as the investigation proceeded. Finally, the chairman summarized the views. 'Let us call on the experts to proceed as they think fit in order to get the maximum information about the contents of the box, bearing in mind, of course, that national security must not be jeopardized. Let the experts, Mr Pande, and the scientists Arul and Laxmanan, prepare a report for us. I hardly need to emphasize the need for complete confidentiality … let none of the findings of the investigation go beyond this committee.'

'When will this committee meet again, sir?' Harisharan forgot that Raj Nath and not he was the convener.

'The committee will meet when the experts prepare their first report.' The chairman looked at his watch in clear indication that the meeting was over.

10 The Fortress

'Umi … Umi …'

As usual, the long distance call had a lot of static, but Urmila managed to make out that it was Laxman at the other end.

'Yes, Urmila here. I bet your stay is extended further.'

'Yes, my angel, extended indefinitely. That's why I am arranging for you to come here to stay. Pack and be ready to travel on the Karnataka Express on Tuesday. Your ticket is being arranged and will be delivered to you. Be sure to bring all you need for a stay of several months.'

'Including your stone god?' Urmila asked jokingly. She was elated at the prospect of being with Laxman again.

The 'stone god' was not an idol. Laxman was not a believer in Hindu rituals, but he loved idlis and dosas. Realizing this, Urmila's mother had given her the traditional stone grinder to prepare the dough for these dishes. Urmila recalled Laxman's vehement veto when she proposed bringing it all the way from Tanjore to Hyderabad. Always a light traveller, Laxman could not tolerate carrying that huge block of stone. Urmila on the other hand did not wish to leave her mother's gift behind. This led to their first quarrel after the wedding, and Laxman thought that Urmila had conceded him victory. However, on reaching their home in Hyderabad, when Laxman opened all the boxes he was astonished to find the stone block peering at him from Urmila's steel trunk. Rather than carry it separately, Urmila had smuggled it in. Only then did Laxman realize why the porters who carried that trunk in Tanjore and Hyderabad had demanded extra tips.

However, once the grinder was placed in Urmila's kitchen, it more than justified its transport and soon attained the status of a stone god in Laxman's gastronomic view.

As Urmila pulled out her famous steel box from under the bed, she speculated about the contents of that other large container, in Gauribidnur.

Karl Shulz, too, was speculating. He never wasted time, and on his first night in Bangalore he had promptly set off for Gauribidnur.

At ten o'clock that night Pyarelal brought the car to the porch of the hotel as arranged, and Shulz was ready for him. The driver set off as previously instructed.

As the car crossed the city limits, Shulz spoke: 'PL, I agree with you … I won't contact Navin just now. We should not be seen together. But what about you? You have been meeting him in the past. Have you ever been watched?'

'Watched?' Pyarelal was shocked. 'By whom?'

'By the secret service … your CBI or IB.'

Pyarelal laughed and added, 'They are all morons. Had they known anything about us they would have nabbed us by now.'

'Don't be too sure, PL. Of course, it is your business. But let me give you that age old advice that I have found indispensable: don't underrate your enemy. For example, did you realize that, as we came out of the hotel, there was a person watching us?'

'Rubbish, Karl! You have become paranoid about our secret service. There are always a lot of hangers-on in our five star hotels ...' Pyarelal suddenly broke off with an exclamation as a car whizzed past in the opposite direction.

'What is the matter?' Shulz asked.

'That was Navin's Jaguar going past. He must have got some news for us since he was heading for Bangalore.'

Shulz cursed under his breath and commented, 'PL, your whole organization is amateurish! Otherwise this lack of coordination would not have happened. Evidently you cannot communicate with Navin inside that Science Centre.'

Pyarelal was silent. Shulz had hit upon the nail well and truly. However, after a while he added lamely, 'Karl, you will soon see how well fortified that place is. See if your professional brain can find a way of ...'

Shulz slapped him on the back genially and added, 'My friend, don't take it to heart! Stop the car about a kilometre from the Science Centre. I will get out and walk.'

The car pulled off the highway behind a clump of thick bushes. Shulz got out and walked on swiftly but noiselessly into the night. As he disappeared into the darkness, Pyarelal wondered what it would be like to encounter this huge figure in the dark as an adversary. He shuddered at the thought.

A ten-foot high brick wall mounted with broken glass surrounded the Science Centre. Moreover, an electrified wire fence had been erected on top of the wall as a further precaution. There must also be burglar alarms suitably hidden, Shulz surmised as he took it all in. He directed his steps to a large peepal tree not far from the fence. Noiselessly, and with a speed that belied his bulk, he climbed halfway up the tree. He was hidden by the foliage but could get a clear view of what lay beyond the wall. The view was barely visible to human eyes, but it was clear enough for the infrared camera that Shulz carried with him. He rapidly took a large number of photographs.

From what little Shulz could make out with his own eyes, he was at the eastern corner of the centre while the excavated portion lay to the south. Not far from him stood what was evidently the laboratory building cum office block. The adjoining barracks to the north were presumably used as housing. Shulz uttered a grunt of satisfaction when he discerned an empty plot to the west, no doubt set aside for a garden.

'Aren't the fortifications formidable?' Pyarelal asked as Shulz returned an hour later.

'Indeed, yes. But I have managed to get prisoners out of jails guarded even more meticulously.' Pyarelal knew that Shulz was stating a fact that the police of a dozen countries could corroborate.

As the car made its way back to Bangalore, Shulz's brain was already formulating plans for penetrating that barrier, if not physically, at least information-wise. Like Houdini, he treated fortifications as a challenge and invariably found ways of getting through them. He estimated (and he was always conservative in his estimates) that within a week he would be able to establish a channel of communication with Navin.

Of course, Shulz thought grimly, Navin must play his part.

Major Samant was unscrambling the telexes from Delhi. He had warned the telex operator that garbled messages from Delhi did not mean that his machine was out of order, and that he need not try to read sense into the jumble of words.

There were five separate telexes on his desk. The key to them was in the secret drawer of his innocent looking desk, specially brought from Delhi. Even with the key, each telex by itself would have made no sense. But when all five were brought together in a certain order, they became clear. Their message drew out a whistle from the phlegmatic man who read them.

11 The Suspect

March 2

Dear Lalitha,

My first letter since the wedding! I realized with a shock that more than a year has passed. You may be wondering if your school friend has forgotten you. Well, I must confess 'guilty'. Although ours was an arranged marriage, Laxman and I have been very happy—happy enough to forget all others!

Remember how you and all our group of friends teased Laxman on the wedding day? Perhaps his shyness on that occasion misled you (and me, too!) into thinking that I was spliced to the prototype absent-minded scientist. No way!

For all his international reputation as a scientist, Laxman is a down-to-earth man who cares about others. He likes to enjoy life—work permitting of course. Others had warned me that he would ignore me because of his work. Until recently I could confidently assert that they were wrong.

But recently, Lalitha, things have changed. You will notice that my letter is not from Hyderabad. We are at Gauribidnur—a small, quiet place to the east of Bangalore. I know your geography was weak in school (remember how our Tope Miss made you locate places on the map?) but try and find out where Gauribidnur is!

Laxman is deeply engrossed in the contents of that wretched container found here. You must have read about it a few weeks ago—how it was found accidentally. Apparently, the contents are fantastic—not just for archaeologists whose minds are in the past, but also for my dear husband who is always thinking of the future.

That's the trouble. Laxman is so engrossed in making sense out of it all that he has forgotten me. He goes out early in the morning, returns late at night, sometimes not at all. He won't disclose a word of what it is all about. It is supposed to be highly classified.

Classified! Secret! Security! These are the operative words where we live, fortified by barbed wire, high walls and armed guards. Major Samant, who supervises all the security arrangements, is a real tyrant. Of course, he is polite and all that—but hard as a nail. Even this letter must go through censorship. But let him read what I frankly think of him. No doubt he will smile and let it through.

We are not let out of this compound except on rare occasions. Laxman has promised to take me one evening to Bangalore. He has a special pass. But at present that is like a politician's promise before elections.

Meanwhile, I must continue in these barracks, the life of a neglected wife, like my namesake in the Ramayana. I am the only housewife around. The other scientists have not brought their families because they have school-going children. There are a few other women here, secretaries, lab assistants and computer programmers. We meet on occasions to gossip …

Laxman has provided me with my veena so that I may continue to practice. But there is no fun playing music if you have nobody to play to. And, of course, I have books to read.

Do write. About yourself and the free outside world you live in. Your letters (if they get through our Major's inspection!) will be great morale-boosters.

My regards to Dr Jayaraman.

Yours, as ever,

Urmila

'Umi, how about going to Bangalore for dinner tonight?' Urmila could not believe her ears. True, Laxman looked very pleased with himself and had even admired her idlis. But this invitation seemed unbelievable in present circumstances.

'Could you repeat what you said? I am sure I did not hear you correctly.'

Laxman's face wore an embarrassed smile. 'I think you heard correctly! This evening, the first stage of my work here will be over and I am already in a mood to celebrate … with you as my honoured guest. For, with thou beside me in this wilderness …'

'You need not wax poetic, Laxman. You know I have done nothing to help you in your work.'

'On the contrary, Umi—it is because of your self-effacement that I have been able to put in so much work. Don't think for a moment that I have not noticed your loneliness here. But I am helpless, for I cannot rest until I get to the bottom of this strange business.'

Urmila gently pulled the unruly locks on Laxman's head—locks to which she had become attached right from their first day together.

'Laxman, one year's training is enough even for a muddle-headed one like me to discover what it is like being married to a scientist ... As for tonight, it's a date! And I won't allow you to back out, come what may.'

So they arranged their programme. Laxman would be free by five in the afternoon, when they would get a car to go to Bangalore. He had already arranged with Major Samant for a staff car to be at their disposal. He left Urmila in a state of sweet anticipation as he walked over to his office.

Yes, today he would put finishing touches to his highly classified report on the container. Although Arul and Navin had helped him, the lion's share of the work so far had been his, and, of course, that of his assistant, the VAX computer at the Science Centre. Nobody knew what his report contained; he had typed it himself on the word processor. It would take him one hour to go over those hundred pages yet once more, before releasing it for discussion with Arul and Navin. And then it would go to their high level committee in Delhi.

'Beep ... Be...ee...ep...', his intercom was calling.

'Laxman here', he responded laconically.

'This is Major Samant', the Major's crisp voice crackled over the intercom. 'Can you come to the coconut grove within ten minutes please? And not a word to anyone that you are meeting me there.'

Laxman was puzzled and vaguely uneasy. He disliked intrigues—in fact, he disliked any interference with this chalked out programme. Why was Samant so mysterious?

The coconut grove was a pleasant spot in the grounds, usually patronized by the staff when they relaxed after lunch. As Laxman made his way there he wondered if Samant had chosen it because it was out of sight of someone working in the main building. Samant was waiting for him and within a minute they were joined by Arul. Presumably Navin would also come, thought Laxman. But the Major began to speak rapidly and in a low voice.

'I have called this meeting to discuss a delicate matter.'

'Shouldn't we wait for Navin?' Arul asked.

The Major looked disturbed at this question. He coughed as if to buy time before speaking. Finally, he blurted, 'The matter concerns him ... he must not know of this meeting.'

As Arul and Laxman waited, he continued, 'You know, I am from the Intelligence Bureau. At the I.B. we have files on all three of you.'

'Big Brother is watching you!' mused Laxman to himself. Arul, however, burst out in anger, 'Are we in a police state?'

'No, Dr Arul.' The Major had recovered his poise. 'We have to do these things for national security. These files remain inactive until we feel that something in a particular file is a potential threat to the nation. I may assure you—although strictly speaking I shouldn't—that the files on both of you are clear and hence inactive.'

He would have said the same even if the situation were otherwise, thought Laxman. However, what the major did not say was more significant. Arul came to the point.

'What about Navin?' he asked. Major Samant shook his head.

'I am afraid Dr. Navin does not fall in that class. Well known as he is amongst the international community of archaeologists, he has also been subject to … ah … illegal pressures to which he has succumbed. He has misused his position and knowledge … we know that he has been responsible for the unauthorized shipping of some valuable pieces out of this country.'

'You mean he is a smuggler'. Arul liked to call a spade a spade.

'And if so, how have you let him be at large?' This was Laxman's question, to which the Major next replied.

'Because of Dr Navin's expertise, we need him here. But more importantly, we are after bigger fish. Interpol is in touch with us and we have decided to wait for a while. We will act in good time … Have you seen this man, by the way?'

The Major suddenly took out a photograph. He was watching the reactions on their faces as they both studied it. Evidently the face was new to them.

'Who is he?' Arul asked.

'A man of many names!' The Major elucidated: 'He has been responsible for many acts of smuggling important documents, archaeological remains, valuable gems. Several countries want him, not just for smuggling, but also for kidnapping and murder … Well, gentlemen, this man has recently been seen in Bangalore. As Karl Shulz, he is staying at the Royal Manor. If he is around it means that things are getting hot.'

'But what makes you link his presence with our project here?' Arul asked. Major Samant pulled out a photograph with another face, which again was unfamiliar to them both.

'This man is Pyarelal. Runs a business in Delhi and has been connected with Dr Navin in some of his shady deals … Well, Pyarelal has been seen with Shulz in one of the Bangalore restaurants. And, as you know, Dr Navin has also been visiting Bangalore.'

'Well, if visiting Bangalore is a crime, then I am about to commit one today.' Laxman smiled in spite of the gravity of the matter under discussion.

The Major also smiled, but added, 'No sir! But Dr Navin was seen with Pyarelal at a Bangalore cabaret last week.'

'It's true that Navin visited the cabaret', Arul recalled. 'He was relating some racy stories in the canteen the next day.'

'Then, yesterday I discovered something more serious. Come, let me show you.' The Major took them to a tiny patch where a cactus garden was coming up. Large stones were placed besides blooming cacti. Major Samant picked one up. There was a hole underneath.

Probably a snake's abode, thought Laxman. But the Major shoved his hand in and took out a tiny, dark object.

'An ingenious transmitter of foreign make! With this, Dr Navin doesn't need to go to Bangalore. He can send messages from right here ... This little toy has a range of ten kilometres.'

The Major replaced the 'toy' in the hole and covered it as before.

'But why don't you confiscate this dangerous thing?' Arul asked, shocked.

'Dr Arul, espionage never goes along the direct routes that are so common to you scientists.' The Major's face was expressionless as he proceeded. 'We have put our own little bug on that toy. So we will get to know exactly who sends out messages and what information is being leaked. So far nothing of value has leaked out ... But I need hardly add, please be cautious and on guard with Dr Navin. He must not suspect.'

'But what about my secret report?' Laxman asked. 'I have to show it to him.'

'With your word processor, I suggest that you give Dr Navin a doctored version, while you and Dr Arul keep the correct one.'

'It's difficult', Laxman said, worry showing on his face. He was trying to recall just how much of the report he had already discussed with Navin.

'Difficult, but not impossible and, under the circumstances, absolutely essential.' The Major was hardly being helpful.

'And we will have to act as if we don't suspect', Arul added. 'Well, we will do our best.'

But it was going to be very tricky, they both felt.

'So this is the Royal Manor?' Urmila asked.

'She was visiting a five-star hotel for the first time and was suitably impressed by the imposing architecture. Laxman had chosen to celebrate his completion of the report in the grand manner and overruled Urmila's suggestions of cheaper restaurants. He also felt that she deserved only the best. Before reaching the hotel, they had visited the shops on Brigade Road—for at Gauribidnur nothing at all special was available. By nine o'clock, they were seated in the restaurant and Urmila was studying the menu. From where he sat Laxman got a good view of the entrance.

'You know, Umi, the best suite in this hotel was once called the Waterloo suite.' Laxman loved telling a tale he had heard from a Bangalore friend.

'Waterloo? After the famous battle?' Urmila queried. The menu was proving incomprehensible to her, so she gave up reading it.

'I'm glad you still remember some history. Well, this suite is reserved for VVIPs when they visit Bangalore. And guess who came to occupy it one day?'

The question evidently did not call for any answer. In any case Urmila did not have one.

'The President of France!' exclaimed Laxman. 'So they had to change the name of the suite at the last moment.'

Urmila laughed, more because she was happy to see her husband relaxed than at the anecdote. The head waiter had meanwhile materialized. Laxman gave his order and left Urmila to work out with the waiter's assistance what she wanted for herself. He was idly studying the cross section of Bangalore gentry walking through the door when, suddenly, he stiffened. A tall, well built white man had just entered and was being shown his table. There was no mistaking him. Laxman nevertheless got as close a look as possible. Yes, it was the man in the photograph he had seen that morning and whom Samant had described …

Karl Shulz.

Guru

1 The Trojan Horse

'Gentlemen, let us open the packet given to us. Like you I am also eager to see what is inside.'

The Minister then proceeded to tear open the manila cover marked 'secret'. The meeting of the 'Container Committee' was convened in Shastri Bhavan this time. Great care had been taken to ensure the utmost secrecy.

'Before the members start reading this report, I would like to make a submission, with the permission of the chair', Laxman spoke up.

'Of course, Dr Laxmanan, go ahead.'

'What you are now going to read has been checked by Dr Navin and Dr Arul. You may be tempted to think so, but let me assure you that nothing is exaggerated. It's a factual account. Even so, I have taken the liberty of not mentioning certain portions that are at present known only to me. They are so sensational that I do not wish to disclose them to this committee without a green signal from the Honourable Minister in the chair. I will be happy to present this additional material to you, sir, after this meeting.' He bowed to the chair and looked at Samant who winked in approval. He had already briefed the Minister about this before the meeting.

'True! A committee with a membership of three or more can never keep a secret—so says a Sanskrit proverb.' Probir Ganguly gave his support. One member, however, felt otherwise.

'With all due respect to Dr Laxmanan's point of view, sir, I wish to state my opposition', Harisharan said. 'If he cannot take us into confidence, we would prefer to withdraw from the committee.'

'Harisharanji! You misunderstand what Dr Laxmanan is trying to say. As we all heard, he has left it to me to decide what can be disclosed and what cannot. And I propose to reserve my judgement until we have gone through this report ... Gentlemen, I give you forty-five minutes to make a rapid reading.' The Minister closed the discussion with a finality that only a seasoned chariman can bring to bear on a committee discussion.

There was pin drop silence for three quarters of an hour, broken only by occasional exclamations of surprise or sharp intakes of breath.

Raj Nath was the first to win the reading race. And the effect the report had on him could be judged from the fact that he forgot to smoke his pipe throughout that period. He rectified this omission by releasing his reaction through a hearty puff.

'Yes, Dr Raj Nath!' the chairman said some five minutes later when everybody had finished.

'Had I been unaware of the circumstances, I would have called this report an exciting piece of science fiction.' Sir, I am not competent to comment on all aspects of the report. I will confine myself to the field that interests me most.

'As we are all aware, Cray is the fastest and biggest supercomputer at present.' Its mode of operation is sequential. There are attempts to branch out into what is called parallel processing. In due course we may hope to surpass the capabilities of Cray by suitable combinations of sequential and parallel computing.

'However, what this report contains is a blueprint for a computer that makes all our present efforts sound primitive. Instead of electronics, it uses photonics. That is, the processing of information in the computer is not done through electrons as in the computers we know today, but through particles of light, the photons. For us a photonic computer is a dream for the future. For those who left the container behind, it was a demonstrated reality. If we can reproduce that achievement, we will have jumped across several steps in computer technology.'

'I agree that it is an important matter and we will take a decision on it today. I am sure my colleague the Minister of Science is equally excited by this possibility.' The chairman looked at the Minister of Science sitting across the table.

'Absolutely! We must take Dr Raj Nath's suggestion very seriously', the M.O.S. replied.

'However, if I may raise a wider question, I would like to ask the experts for their overall assessment of the role of this container as an indicator of a past civilization. Dr Pande, the report does not go into the details of how you deciphered the alien language. Could you enlighten us on this issue?' The chairman put the ball squarely in Navin's court. Navin, of course, was well prepared.

'Sir, first let me say something about the age of the container. Radiocarbon dating puts it at around twenty thousand years. Comparing the proportion of carbon fourteen in the container with other known archaeological remains, we can agree on this estimate', Navin looked at Dr Gupta for support.

'Navin is right,' Dr Gupta spoke in his steady, sure tone. 'Let me explain for the non-experts. Normal carbon found in nature has an atomic weight of twelve; that is, its atom is twelve times heavier than the atom of hydrogen. However, one also finds on this planet a heavier brand of carbon in small proportions. This is carbon fourteen—with an atomic weight of fourteen—which happens to be radioactive. It decays into the lighter species through radiation. We know at what rate it decays. If you take any population of carbon twelve now, half of it will decay in about 5700 years. So by measuring the abundance of this radioactive species in archaeological relics we can estimate their ages.'

'Thanks, Dr Gupta, for explaining this to us. I have been hearing the phrase carbon dating for so many years—only today have I understood its significance. Dr Pande, please continue', the Minister added.

Navin paused briefly to take stock of what he had to say. Everyone waited expectantly.

'Gentlemen, you have read in the report how we managed to open the container. As we examined its contents we began to appreciate how logically those people had arranged them. They started with numbers. We use ten digits, they used eight ... the so-called octal system.'

'Does it mean that they had four fingers in each hand?' Harisharan hazarded what he thought was a clever guess.

'No. They had five fingers in each hand—in fact they were humans just like us. Their preference for the octal system was based entirely on convenience. As you know, the computer uses binary arithmetic in which every number is expressed with just two digits, zero and one. It is the most fundamental of all digital systems, but it requires a large number of digits to express any number. Since eight is expressed as $2 \times 2 \times 2$, the octal system can be easily converted to the binary and vice versa. At the same time you don't need a large number of digits to write a number....'

'I will take your word for it, Dr Pande, although I must confess that much of this is going above my head.' The Minister smiled as he made the confession.

'Mine too, for I never liked maths and was glad to give it up at the first opportunity', Navin smiled in return and added, 'but even I could make out that in the theory of numbers these fellows were leagues ahead of us. Perhaps Dr Arul could comment on this.'

'What Dr Navin has said is, if anything, an understatement. Abstract mathematics can in principle be described entirely in symbols—without words. So it was easier for us to decipher their mathematical writings. Let me tell you that they had solved most of the problems that today's mathematicians consider intractable. For example, the Riemann hypothesis....'

'I don't think we need go into the details, Dr Arul!' The Minister could see the baffled faces around the table. He motioned to Navin to continue.

'Turning to words, these people had prepared a dictionary of sorts in which pictures of concrete objects were given side by side with their names.' Moreover, they had evolved a method of digitizing the letters so that each word could be fed to a computer.

'Even their grammar was expressed in mathematical language! It was therefore easy for me to understand how they constructed sentences....'

'If I may interject, sir', Laxman added, 'Navin is understating again, but this time with respect to his own achievement. It was a pleasure to watch him unravel the grammar and get at the language.'

'Thank you', Navin said with a mock bow. 'From sentences to information was the next step. I am still collecting and documenting the information about the social conditions of those people. I will present a concise report to this committee when I am through. I think Arul should comment on their science.'

At a sign from the chairman, Arul started his piece, which he had carefully rehearsed so as not to tread on sensitive toes. He began: 'I have always had difficulty convincing myself as a scientist that our Vedic ancestors were technologically advanced. For none of the descriptions to date have contained what we call hard core science, that is, laws mathematically worded and data quantified. Evidence in the form of precise figures and construction kits has been conspicuously lacking. Not so for the container people! Their documentation includes all these, from basic physics, chemistry and biology to their technological applications. The report touches on a few examples, but these are by way of being the tip of the iceberg. There is enough information to keep our research labs busy for years.'

'May I say something, sir?'

The Minister looked towards the smokescreen from behind which these words had emerged.

'By all means, Dr Raj Nath.'

'It seems to me, sir, that there is a lot of knowledge in there waiting to be unravelled. But the human brain, such as it is today, may take several years to do so unaided ... Unaided that is, by their supercomputer.'

'I get your point, Dr Raj Nath, we need to build this computer as fast as we can. This committee should give it top priority', the chairman observed.

'I endorse it wholeheartedly', added Probir Ganguly. 'We need to take a holistic view—and it leads us to the computer. The computer is needed to understand and interpret all the knowledge buried in the box. Converting that knowledge to useful ends will have far-reaching consequences. Let us build it as a top priority.'

All except two around the table spoke up in support. The chairman, always very perceptive, noted this and decided to probe further.

'Harisharanji, you have been silent!'

Harisharan was waiting for this opportunity. He spoke in his most official tone. 'The idea is indeed worth supporting. But there are several practical difficulties that I must mention. First of all, Plan funds have already been allocated and there seems to be no way of getting funds for this project. Second, the Department of Electronics has recently issued a strict order barring all new computers. Then, of course, there is the question of which ministry will serve as the nodal ministry for it.'

In short, he does not want to be saddled with anything new, thought Laxman. He exchanged glances with Arul, who had also arrived at the same conclusion. But the chairman was undaunted. 'Points all well taken. So far as D.O.E. is concerned, I am sure there will be no problem. The ban is on imported computers. We are making the computer here, not importing it. Regarding money, we all realize that this presents us with an unprecedented opportunity for which we have to make new rules. Just before coming here, the P.M. called me. He is greatly interested and has assured me that there will be no problem of a financial nature. In fact, in view of his interest, let us resolve to request the P.M. that his office will be the nodal ministry.'

'Hear! Hear?' Navin broke out excitedly. He had echoed the mood of the committee. The chairman, however, had to solicit the view of another silent member.

'Professor Kirtikar, you have been quiet so far. May we know what you think about this project?'

Given this question, Kirtikar had no option but to speak out his reservations.

'Mr Chairman, in normal circumstances one would, of course, have unreservedly welcomed the opportunity of constructing this computer.' In fact it would have been silly to let the opportunity pass us by. But, gentlemen, I am reminded of the story of the Trojan Horse. The Trojans under siege found a large mechanical horse left outside their citadel by the Greek army. They took it inside ... and that was their undoing. This story led to the saying "Beware of Greeks bearing gifts". Before we accept the gift of this container we should think twice....

'Why is this civilization that was once so advanced extinct today?' Of course civilizations can be destroyed for two reasons. Natural causes, like earthquakes, volcanic eruptions and ice ages can easily destroy a very advanced civilization. If this one perished for such reasons, I have nothing further to say....

'But what if it was destroyed for man-made reasons? Today we are fairly advanced compared to our ancestors of a century back. But our technologies have brought their own perils, for example, nuclear extinction or large-scale pollution. What if this container has information that is potentially lethal? Would it not be wiser to first find out why those people were destroyed?'

As Kirtikar made this impassioned speech he looked round for reactions. He realized that his caution would fall on deaf ears. Strange, he felt, that the only person on his side in this mixed assembly was the one he least expected. And even Harisharan's support was based on entirely different arguments that would not serve his purpose.

There was a discussion of Kirtikar's point of view, but the committee would not budge from its decision to go ahead. The only concession he got was a rider attached to that decision: 'Great caution should be exercised in applying for practical purposes any information pertaining to or coming from the container.' A concession that he knew very well was merely intended to keep him satisfied.

As the meeting broke up and Laxman followed the Minister to his sanctum sanctorum, Arul took Kirtikar aside.

'When are you leaving for Bombay, sir?'

'By the last flight tonight. Why?'

'Laxman and I want to talk to you quietly … may be for an hour.'

'Then come to INSA at five in the afternoon. I will be through with my meeting by then.'

As he left Shastri Bhavan, Kirtikar wondered what it was all about.

2 The Machine

Navin and Pyarelal were having lunch at the Nalanda, a restaurant of Gautam Hotel near the Pusa Circle. Although a sumptuous buffet had been laid on, the two were more interested in talking in a dimly lit corner.

'I need the manual for the computer … a copy would of course do', Pyarelal said.

'Impossible! Absolutely impossible! As I told you, this information is top secret and Laxman has deposited it with the Minister … I guess the Minister will constitute a task force to oversee the building of the computer. But I won't be on it for sure! My brief henceforth is to compile the history of the container people.' Navin's voice rose in expostulation.

'Take it easy Navinbhai', Pyarelal purred in his peculiar, silky voice. 'Nothing is impossible in this world. In any case, with suitable lubrication even the stiffest joint can be made to move.'

Navin hesitated. He knew what lubrication Pyarelal could apply.

'Karl has gone abroad to find the clientele. Be assured that this time we are talking of really high stakes ... How about one lakh?'

Pyarelal held out a finger. Navin held out both his palms with all fingers opened out.

'You are asking a lot, Navinbhai.'

'As you said, PL, these are high stakes. You are lucky that I have only two hands with only five fingers on each.'

'OK, done! We will celebrate.' Pyarelal called the waiter to serve the drinks. The alacrity with which Pyarelal had accepted his demand made Navin wonder if he had underbid.

As they were celebrating, a young man on a nearby table got up. He then slipped out and went to a public call box in the lobby downstairs, where he dialled a local number and talked for a long time.

Traffic was at its peak on Balhadurshah Zafar Marg as the scooter brought Arul and Laxman through the gates of the Indian National Science Academy. Two pink buildings stood in the grounds and Arul asked the driver to drop them at the building on the right. The beginnings of summer were noticeable, and they were relieved to leave the hot, dusty road and enter the waiting room.

The receptionist informed them in Hindi that the meeting was still going on and that Kirtikar Sahib was inside in the committee room. They waited on a sofa beneath a ceiling fan.

'You have been here before?' It was more a statement than a question. Still, Arul felt like answering.

'Five years ago I came here in April to present myself for an interview on my research work. I had been shortlisted for the INSA Young Scientist Award.'

'Did you get it?'

'I did! But the awards are not given here. They are given by the P.M. or the President at the academy meeting held at the annual congregation of the Science Congress. That year it was in Lucknow ... Ah, here is Professor Kirtikar.'

INSA had been founded in 1934, but it had really come into prominence after independence. It patterned itself on the Royal Society and, with its headquarters moved to Delhi from Calcutta, it served the many official purposes that a national academy is called upon to perform. Kirtikar had come for one of the numerous committees that INSA constituted to conduct its official business.

'Let us go into the committee room—it is air-conditioned', he led them in as other members of the committee walked out with their papers and briefcases.

The committee room was very spacious. Small tables had been joined together to make an oval-shaped ring that could easily have sat fifty. The smaller committees occupied only one corner, as was the case today. When the secretary and typist had cleared all the papers and left, Kirtikar motioned to Arul.

'Arul, the floor is yours!'

'Well, Laxman can describe it better than I can … in fact, I myself don't know most of it.' At this prompting, Laxman produced a sheaf of papers.

'Sir, today I gave some secret documents to the Minister. They basically contain the blueprint for the computer. Sensational as a photonic computer will be, what I have here is even more so. I need your advice on how to handle such a hot matter.'

'Must be pretty hot if you did not want to part with it even for the Minister!' Kirtikar commented drily, looking at the sheaf of papers.

Arul and Laxman looked at each other. Finally, Laxman proceeded further. 'Sir, unfortunately we have a mole in our midst right in the Science Centre … Navin.'

'Navin Pande?' exclaimed Kirtikar. 'A distinguished archaeologist like him? I can't believe it.'

'Unfortunately it's true. Major Samant has a big file on him and he has advised us to be cautious. Which is why I had to use the subterfuge I did at the meeting this morning.' Laxman said. Arul then narrated their encounter with Samant.

'It figures … well Laxman, if I may so address you informally, you did the right thing … but what a shock about Navin! … To come to the business, however. May I read these now?'

'Please do!' Laxman added.

As Kirtikar finished the last page his face clearly reflected an internal turmoil. 'I can't believe it!' he added, still holding on to the bundle.

'Nor did I … but then, with this container I have relaxed my bounds of credibility', Laxman said.

'A von Neumann machine! A blueprint for this fantastic thing right in our hands? Laxman, are you sure it will work?'

'What is a von Neumann machine?' Arul interjected.

Laxman clarified: 'Arul, you have heard of the mathematical genius John von Neumann. Amongst his researches in artificial intelligence during the nineteen-fifties was the notion of a machine—a robot if you like—that can reproduce itself. Von Neumann proved mathematically that such a machine can in principle exist. But from a mathematical construction to technological achievability is a long step. Nobody believes that von Neumann's construction is achievable in the foreseeable future. Not with our present technology.'

'And now you have evidence that these container people had succeeded in making such robots … unbelievable!' Professor Kirtikar said.

'But true! These robots can be made once we have the computer. For the logical maze needed for their construction can only be penetrated with the supercomputer ... But once we make them, they will prove tremendous assets to our technology. These robots played an important role in the lives of their makers. This much is now clear from whatever account we have about the container people.'

Kirtikar was silent for a while. Then he spoke, a thoughtful expression on his face. 'You know my views expressed this morning. A photonic supercomputer is bad enough ... now this further step of a self-replicating robot! My mind simply boggles.'

'I agree, it is a hot potato ... and that's why you have to tell us how to handle it. More so because Navin and his lot will be after it ... We are totally out of our depth', Laxman added.

Kirtikar was pacing up and down the long committee room—his habit whenever in deep thought. Arul motioned Laxman to silence as they waited patiently. They could hear the rumble of traffic on Bahadurshah Zafar Marg. What a contrast between that world and the ideas they were grappling with....

Suddenly Kirtikar stopped. His face had cleared and he now spoke in decisive tones: 'Our strategy, my young friends, must be like this....'

3 The Client

The enormous black limousine was conspicuous even in the prosperous residential neighbourhood in Silicon Valley, the capital of California's computer industry. The chauffeur was impressively dressed with a peaked cap; but more impressive was the sole occupant of the back seat. Had any passer-by managed to look through the darkened window he would have seen the classic, inscrutable oriental face.

As it was, there were no passers-by on the well laid out footpaths. The car glided through Main Street with its fast-food shops, then went out on to the highway in open country.

'Turn left here.' The back seat occupant whispered into the speaking tube. They were approaching a minor crossing, where a dirt track intersected the highway.

'Christ!' Muttered the chauffeur, looking at the sorry state of the track. Hardly suitable for his beautiful vehicle. Fortunately, it did not have to endure the ordeal for long.

'Over there, by that shack with the green roof.' Yamamoto's instructions were precise. The shack stood totally isolated, with no other habitation in sight. An ideal place for a secret rendezvous.

The chauffeur pulled into the small enclosure surrounding the shack. He pressed a button on the dashboard to open a secret drawer from where he pulled out a tiny automatic pistol. Slipping it into his pocket, he got out, went round to the other side and opened the back door.

'I don't think you will need the toy, Jim.' A faint smile crossed Yamamoto's face, but only for an instant.

'Won't do any harm having it around, sir. You never know', replied the chauffeur, studying his surroundings with keen professional eyes.

For James Gibbon had once been a crack FBI agent. He had been slated for quick promotion for his many achievements, but gave it all up to be Yamamoto's personal companion-cum-bodyguard—at a salary several times more than the Federal Government could have offered him in the foreseeable future.

As they approached the shack they noticed a powerful motorcycle parked against the side wall. But for it, they might easily be in the frontier days of the last century.

'He is waiting. You'd better remain here since I have to go in alone', Yamamoto ordered. Jim had already noted movement inside, through the crack in the ramshackled door.

'Sir', winked Jim as he saw Yamamoto lightly tap his Omega wrist watch.

Indeed, people who met Yamamoto often wondered why a proud Japanese man like him used a Swiss watch. None except James Gibbon knew of the ultra efficient Japanese transmitter that the watch concealed. It could summon James to his master's side whenever needed—as two years ago in the seedy Los Angeles suburb. Two ruffians had cornered Yamamoto with demands for whatever cash he had. To those professional muggers this seemingly sedate Japanese man had appeared easy prey. But they were in for a shock. One of them received a painful karate chop while the other had a bullet through his shoulder. How the two events happened simultaneously was a puzzle the two victims were trying to figure out in the hospital subsequently.

Without knocking, Yamamoto pushed the door and went in.

He was momentarily blinded by the darkness within. Then he began to make out faint outlines, thanks to some light coming through chinks in the door and the window.

The room contained two old chairs drawn up near a broken table. Behind stood an enormous man.

'Joseph?' asked Yamamoto.

'Yes. I know you, Dr Chushiro Yamamoto. Let us sit down.' The man took the nearest chair.

'Since you don't care to tell me your family name, let us stick to first names only. Call me Chushiro.' Yamamoto spoke in even tones as he sat down.

'Sure! Now coming to business, Chushiro, you are proud of this new super-computer that your multinational is shortly going to bring out….'

'Justly so, I think. We expect to be recognized as Number One.'

Joseph laughed softly. 'In the fifties we had hand-operated calculators. Suppose a dealer in those machines were to proclaim himself Number One today, how would he fare?' he asked. Yamamoto was nettled by this jibe, but his reply did not display this as he spoke. 'Our technology is not only the latest, it looks to the future.'

Even as he said this, Yamamoto felt uneasy. Was Joseph playing cat and mouse with him … before pouncing to kill? Nobody had done this to him so far and he did not relish the experience. But Joseph soon put him out of his suspense.

'Chushiro, let me assure you that your company will soon find itself in that state. See for yourself how computer technology has marched streets ahead of where you are now.' Joseph produced a sheet of paper from one pocket and a small torch from another.

In the torchlight Yamamoto glanced at the paper and stiffened involuntarily. A Japanese expletive was his response as he completed the reading. Joseph would have been flattered had he known that, for the first time in his life, Yamamoto had panicked, if only momentarily.

'I beg your pardon?' Joseph asked as Yamamoto handed the sheet back.

'Joseph, you have got something out of science fiction, I am afraid. A photonic computer of this capability does not exist. Nor is it likely to materialize in the near future.'

'Exactly what a maker of hand-operated calculators would have said in the fifties about the supercomputers of today … But supposing for a moment that this is hard fact rather than fiction?' Joseph tapped the paper as he spoke.

'That is a very improbable thing to suppose, Joseph. In our corporation we keep tabs on what is happening elsewhere. I am fully aware who our rivals in this game are—what they are up to and what they are capable of. To expect any of them to produce this computer is, well, like supposing that a mouse can kill a tiger.' Yamamoto paused to think, and then added, 'This technology lies several decades into the future—it is a dream to which the likes of us cannot aspire. But I have no time right now to waste on dreams. So unless you have something concrete to offer, let us not waste each other's time.'

Joseph let out a guffaw. 'Well spoken Chushiro! But it is not in my interest to put all my cards on the table. It is a game of poker at which, I am sure, you are very adept.' He watched Yamamoto's dead-pan face looking surrealistic in that dim light. Then he proceeded further: 'I may be bluffing, in which case you don't have to worry. You lose nothing by telling me to beat it. But, on the other hand, if I am sincere, then you lose everything by that action. For I will then go to one of your rivals.'

'But for me to take you seriously you have to disclose something more. None of my rivals will take you seriously either, merely on the basis of this.' Yamamoto pointed to the paper in his hand.

Joseph shook his head. 'Right now, I have nothing more to give. But think: would I have gone to the trouble of coming to this country where I am on the wanted list and meeting you in this god-forsaken spot just for the fun of it? I will provide further proofs, but only in stages.'

'Stages? What do you mean?' Yamamoto asked, although with his characteristic shrewdness he had guessed the answer.

Joseph produced a tiny slip with a number written on it. He turned the torchlight on it as he said: 'Since you are gambling, Chushiro, you have to put some money down on the assumption that I am genuine.'

'A Swiss bank account in Zurich, presumably? How much?'

'The amount is written on the other side.'

As Yamamoto turned the paper over and read it, the same Japanese expletive came from him again.

'The amount is nothing to what your firm will make once you get hold of this technology.'

Genuine or not, Joseph was correct on this count. There was no question that whoever could make a computer of these capabilities would call the tune on the world market.

'What do I get in return for this amount?'

'The source of this computer. And, also, if you give me two problems which are beyond the scope of your best computer in the matter of complexity of logic, capacity of storage space and time of computation … I will undertake to have them solved for you.'

'You did ask me to bring two problems … well, here they are.' Yamamoto handed him two printouts and added, 'Our experts say that these cannot be solved until the advent of the next generation of computers.'

'You will get the solution within a week of the date you deposit the first instalment', Joseph replied.

'First instalment?' Yamamoto asked.

'Of course! Surely you agree that knowledge has its price?'

Yamamoto pondered a while. Then he spoke with a tone of finality. 'Joseph, I started earning money by delivering morning papers. It was because I took gambles which paid off that I reached where I am today. I have lost a few gambles … but I don't recall a case where I was cheated and the person responsible for it got away with it.'

In spite of his hefty size Joseph felt a momentary twinge of fear at this warning. Pulling himself together, he asked:

'So, your decision, Chushiro?'

'I accept your terms, Joseph.'

'Let's shake hands over the deal—honour among thieves, eh?'

If Yamamoto disliked being called a thief, his face did not register it. He shook hands with Joseph.

It was then that Joseph realized how firm Yamamoto's grip was.

Joseph stepped out after Yamamoto's limousine disappeared from the scene. Having put on his helmet and goggles he set off on his motorcycle. Soon he was racing along Interstate 5 and headed for Los Angeles.

A half hour's ride along the freeway brought him to a rest area where he had a quick meal and then approached a call box. His telephone card carried the name Joseph Burridge. As he dialled an international call he looked at his watch and smiled softly. It was getting close to four in the afternoon.

'Who is it?' asked an irritable voice after the phone had rung several times.

'Good morning PL! Karl here. Don't you Indians get up early in the morning? Get up, man! It's bright and sunny over here.'

'Surely you are not calling with the intention of providing a weather bulletin?' Pyarelal's voice was quite friendly now.

'No. Just to report. Number One is interested. The Jap will pay the first instalment.'

'Good! Well done! How much did you ask?'

'That is none of your business, old man. They will pay more for further information, of course. I will pay you what we agreed upon. But please arrange for more details on the computer.'

'Will do. Bye.'

As Pyarelal replaced the receiver, a click could be heard somewhere down the line and a tape recorder was automatically turned off.

4 The Wiretapper

'Go on, Mr Singh. You must have something really important to have called me here all the way from Gauribidnur.' Major Samant eased himself on to the sofa as he spoke.

Kamala Prasad Singh had a first class M.Sc with statistics, but had somehow got diverted to the police department. As a police officer he was something of a misfit, being too urbane to lead riot squads against unruly mobs. So he was diverted again, via the C.I.D. to the Intelligence Bureau, where he had found his niche.

'Take it easy, Samantji! Here, I will play a tape for you', Singh said in his polite Hindi.

Singh was a lover of music and was known for his stock of tapes and LPs of classical Indian music. But the tape he played on this occasion was somewhat different. Still, Major Samant liked it so much that he called for a replay, which he listened to carefully.

'Technically illegal of course—but we have to do this from time to time', said Singh almost apologetically.

'I recognized Pyarelal's voice right away. The caller's voice was not so clear, but fortunately he identified himself … Good you called me, Mr Singh.' Samant was clearly very excited.

'On rare occasions we catch gems like this out of a whole lot of junk', Singh said modestly.

'Can you work out where the call came from?'

Singh was hoping this question would be asked. For this is where his analytical mind revelled.

'This call came when it was four-thirty in the morning, Indian Standard Time. That makes it midnight in London and Western Europe because it is summer, seven in the evening in New York and four in the afternoon on the West Coast of the United States.'

Singh paused to let all this sink in and then continued.

'Of course, in summer it can be bright and sunny even at seven p.m. in New York. So I got the Met office to give me satellite weather charts over the United States. Fortunately for me, there were low pressure areas almost all over the country.'

Singh fished out a sheaf of maps from his desk and placed them before the Major who immediately said 'Right you are. All states except California and Arizona seem to be under heavy cloud cover.'

'Taking that with reference to the computer, I homed in on Silicon Valley in California. Our friend was negotiating with somebody high up in Number One.'

Major Samant was still peering over the maps as Singh produced lemonade from his desktop fridge. He was a strict teetotaler.

'There can be no doubt about which is Number One, especially after reference to "the Jap" in the conversation'.

'Who is the Jap?' asked Samant.

'Chushiro Yamamoto, born in Osaka, age forty-five, Ph.D. from Cambridge. Got mentioned for his exceptional thesis on computer hardware. Had many lucrative jobs waiting for him in the USA and Japan but preferred to do post-doctoral research for a few years, where he went from strength to strength. Then he started his own company. Judged the market right when the demand for personal computers rose and rose. Then sold out just before the downhill started and joined a top class multinational in an important policy-making post, where he has been since—now a force to reckon with. I have managed to get a couple of photographs of him, too.' He placed a few magazines on the small table in front of Major Samant.

'You certainly have done your homework, Mr Singh.' Major Samant glanced through the article in a leading computer journal from the West, wherein Yamamoto's achievements were highlighted. Singh grinned with satisfaction.

'Thank you, Major! I might further add that his company has given Yamamoto *carte blanche* to use whatever methods he deems fit to bring it to Number One place and keep it there. Industrial espionage is one of his specialities.'

'All is fair in love and war, eh?' the Major said jokingly. But already his cautious mind was estimating the consequences. That Yamamoto was interested in the Gauribidnur computer was bad enough. His joining forces with an international criminal like Karl Shulz added another dimension to his worry.

'It is high time we took in Navin Pande and Pyarelal. We simply cannot let this leak go on any more', he said in decisive tones.

Singh's deprecating cough told him that such precipitate action was unwelcome.

'Hardly what I expected of you, Major.' Singh elaborated, 'What will that do? May be those two will get a few years in jail. But what about the big fish, Shulz? Will you let him go? Remember, he will carry on with someone else instead.'

'I will have to do so Mr Singh, much though I regret that alternative. My main concern is to protect our project at Gauribidnur. I cannot compromise on that.'

'Suppose you catch Shulz without compromising your project?' Singh asked.

'How? In what way?' the Major asked hopefully—for he could see that his companion's face was shining with excitement. From past experience he knew that Singh had got one of his brain waves. And waves from such a brain as his could not be ignored.

'Let me explain', Singh added, replenishing his glass with more iced lemonade.

'Time to go, Runa', Navin said, looking at the bedside clock.

'Not yet, darling … can't you spare even a single night for your Runa? For two months I have missed you', Runa pulled him back to bed and gave him a long hug.

Normally, this should have been enough incentive to keep him back—but not today. Navin had a flight to catch—the first flight to Bombay. He gently extricated himself and got up.

'Runa, love, much though I hate it, I have to leave. Otherwise I will miss this wretched flight … But I promise you another rendezvous within a month.'

'A tall promise! And no doubt your bodyguard will be there too? I bet he is waiting outside.'

Navin looked out of the window. A solitary figure in plain clothes could be made out vaguely, sitting near the gatepost of the apartment building. It was indeed hard to do much under this constant shadow, for which he heartily cursed the absent Major.

'Runa, you and I have to put up with this for may be a year, until I unravel all the treasure in that container. It's an archaeologist's dream, Runa. And it's a challenge to me.'

As Navin rapidly dressed, Runa made another proposition. 'How about taking me to that god-forsaken place?' Navin laughed at the suggestion.

'You! In Gauribidnur! What would a party girl like you do in that monastery. You will want to quit in two days!'

'I know why you don't want me there, Navin. You have acquired another friend.'

Navin suddenly turned serious. His moment had come. Like a magician, he produced a jewel box from his pocket and opened it to take out a ring.

'Give me your second finger of the left hand.' Even before Runa knew what he meant he had slipped the ring on it. She stared at the brilliant diamond, half dazed.

'What is this for?' she asked, impressed but hardly able to believe the implications.

'I thought even a muddle-headed one like you would know! It happens to be an engagement ring … Runa will you marry me?'

'I can't believe it! A casanova like you asking that question?'

'Well, what is the answer, my love?' asked Navin, taking her in his arms.

Before Runa could answer, however, the door bell rang, followed by loud, decisive knocks.

'Curse it', muttered Navin as Runa went to open the door. It was still too early for the taxi to the airport that he had ordered.

'Who is it?' he heard Runa ask before opening the door. Then he saw her look through the peephole and open the door in some trepidation.

He caught a glimpse of Hajarimal, his 'bodyguard'. But the figure behind him made Navin stiffen.

Major Samant was standing at the doorway in full uniform.

5 The Computer

Dear Urmila

Our correspondence has been going on for several months. You may accuse me of being lazy—you would be justified too, for I write one letter after receiving four of yours! Nor are my letters ever as long as yours.

The fact is that I simply don't have time. Unlike you (consigned to a quiet life in the wilderness), I am in the thick of a busy housewife's routine. With just Jayaram and myself, a family of two, what keeps me so busy? You may well ask! The answer is 'a steady unending stream of guests'.

Jay likes people, whether they are artists, musicians, litterateurs, journalists, scientists or academics—even politicians of all parties. Jay loves to chat with them all for hours on end while poor I have to supply food to keep them going. Tonight he is bringing two film actors for an overnight stay. I have just finished cooking dinner and am snatching a few minutes to scribble this to you. Well, Urmila, I have important news to report. The D-Day is around December 20. This means I have to slow down and employ some regular domestic help in due course. Aren't you lucky that you have a gardener and a woman who cooks? You tell me that they are the spies employed by that Major of yours. Well, I have no use for a gardener in our eighth floor flat but could certainly do with a cook (so I envy you!). Even Jay has registered the news in his busy mind and is talking of reducing the flow of guests. I will believe that when it really happens. Meanwhile I have just sighted Jay's car, presumably containing the dreaded guests.

So I will close for now. My best regards to Laxmanan (when you see him!) and to you—I will also add those from Jay, whom I can now hear at the door.
  Lalitha

'When you see him!' How right she is, thought Urmila as she read the last paragraph. For she rarely saw Laxman these days. He would come back late at night and, on his strict instructions, she had stopped waiting for him for dinner. She would be fast asleep when he came and helped himself to whatever was kept on the table. And by the time she woke up he would be gone. On some nights he would grab a stale sandwich from the canteen and sleep on the bench in the lab.

Urmila sighed and looked out. Jeevan, the gardener was busy weeding the lawn. But, Urmila knew, he was alert for anything unusual. Her other shadow, Rakhee, was putting finishing touches to a lunch of rice and sambar. Though Urmila did not need or want a cook, she had one courtesy of Major Samant—in case something unusual happened.

It did! Masculine arms were round her and she was lifted and whirled round the room. Her instinct to cry out was suppressed by surprise at finding that the arms belonged to her absconding husband.

'Let go, Laxman, let go! What will Rakhee think?' Urmila managed to extricate herself with great difficulty.

'I bet she has been trained to ignore such demonstrations' said Laxman who looked obviously excited. 'Umi, I have finally done it!' he added.

'What have you done, Laxman?' Urmila asked. Some of the excitement had already rubbed off on her.

'I have, shall we say, completed the jigsaw puzzle I was trying to put together for so long.'

'What puzzle?'

'Ah, there you have me. I shouldn't tell you really—but to hell with security—I will. Umi, the computer is working. It's fantastic, decades ahead of the finest supercomputer on this planet. But having said that, I must shut up. And, Umi, keep it to yourself.'

'Congratulations, darling!' Urmila could not decide whether she was more pleased at Laxman's success or because there was now the likelihood of his finding more time for her.

The excitement had all begun with the first trial run, when Arul had produced a programme for the computer....

'What programme is this?' asked Laxman.

'Back in 1976 two mathematicians Appel and Haaken used this programme to solve the long-standing four-colour problem', said Arul as he handed the floppy disc to Laxman. As Laxman went over to put the disc into the new computer, Arul had second thoughts on the matter. He added, 'Laxman, you wanted a long-running programme; but let me warn you that this one took nearly 1200 hours on IBM-360 ... Do you want something more modest to start with?'

'No Arul, I will take it as a challenge. I could have tried shorter and simpler problems—but they won't give a real indication of its capability. Let me first put the disc in to translate the programme language to one which this computer understands.'

'What is this business of translating languages?' asked Navin.

'This computer has been designed as per specifications totally alien to us. Even our own computer systems in the initial stages did not readily adopt a common system of programming. So one had to design a translator to change one programming language to another. I designed one that this computer understands. That is why it took me so long to make this computer work. Of course, in due course we must get used to the computer's own language:

then this delay won't occur', Laxman said as he anxiously watched the flashing light. It would change over to a steady red light if the computer did not understand some instruction in the programme.

'Bravo!' shouted Laxman with delight as the flashing light stopped and a series of beeps came out. The indicator light had turned to green. The computer had understood all the instructions. Laxman took out the floppy disc and pressed another button.

'The computer will now examine the programme for its internal logic', he added.

'While it is doing so, can you explain to me the four-colour problem?' Navin asked Arul.

'It is one of these conjectures which are simple to state but difficult to settle one way or another', said Arul. 'Suppose on a geographical map you want to depict countries by different colours. The condition is: adjacent countries, that is, those sharing a boundary, must have different colours. What is the least number of colours needed to paint the map?'

'Wait a minute! That looks simple enough', Navin went to a desk and pulled out a paper. He drew several maps to check his guess and finally said. 'Four colours ought to be enough'.

'Right! That is the four-colour conjecture. But can you prove it? Can you show that, no matter how you draw the map, four colours will suffice?' Arul asked.

'Well … now you are asking! May be there is some map cleverly drawn that might need five colours … let me try a little more.' Navin started drawing a few more maps.

'You can save yourself the trouble! People have been at it for over a century. Some thought that they had proved the conjecture but their proofs contained fallacies. Nor could anyone find a map that needed at least five colours.'

'There must be maps and maps. Unless one exhausted all kinds of maps that can be drawn, one cannot really settle this issue', Navin said.

'Well, as a rule mathematicians look for a general proof that covers all cases without having to specifically deal with individual examples. Take the Pythagoras theorem about right-angle triangles. You don't need to draw each and every right-angle triangle to show that the theorem works. You construct a general proof', Arul explained.

'What a proof! It was enough to turn me away from mathematics at school', Navin confessed with a wry smile.

'A simple proof it was, if only you had good teachers to explain it to you, Navin! For the four-colour conjecture, no compact analytical proof was forthcoming. So Appel and Haaken did something that pure mathematicians rarely resort to. They went to a computer for help. They had already classified maps into different categories depending on their complexity. But there were far too many categories and far too many maps in each category to be within the range of a human brain to verify. As I said earlier, even a fast computer could not settle the matter easily.' Arul glanced anxiously at Laxman who was watching the console intently.

'So what Laxman is doing to his computer is like asking a toddler to break an Olympic record.' Navin now began to appreciate the immensity of the problem. Would they have to wait for fifty days to get the answer?

'Good ... good ... good! The computer has accepted the programme' Laxman shouted suddenly. 'Shall we start it?' He pressed the 'execute' button without waiting for their concurrence.

'How long should we wait for an answer, Laxman? Hope not 1200 hours', Navin asked jokingly.

'I estimate it to be less than an hour—at a conservative guess', Laxman said.

'Come, let us have a cup of tea while we wait.' Arul dragged the other two, Laxman especially, to the canteen. He felt that Laxman could hardly bear to wait there in suspense.

Even in the canteen Laxman could not relax. He kept thinking of the computer he had put together according to the instructions found in the container. It had been a long haul. For this he had had to commission components from R & D labs which had never encountered this type of technology before. It had meant trial and error until the specifications laid down so clearly in those instructions were met. And all this had been done in total secrecy. One lab did not know what the others were doing. Within the same lab, different scientists were assigned non-overlapping jobs, all classified.

Would it work? Laxman could not imagine the consequences of a failure.

'Shall we go back and have a peep?' he asked suddenly.

'It's hardly been fifteen minutes ... you wanted an hour', Arul reminded him.

'All the same, let me check and see if it is still working.' Laxman did not wait for the others to finish their coffee. He pushed aside his cup and got up.

The computer was not working. His heart sank as he saw the red light. How would he face Arul and Navin now? He could hear their steps behind him....

'Your toddler appears to have collapsed!' Navin's voice was soon heard from the doorway. He too had seen the red light. What would Arul say? That he had been too ambitious in asking for such a long programme for a first trial?

Surprisingly, Arul was silent. He was staring at the monitor located a few feet away.

'Laxman, come, have a look! I can't believe it', he finally blurted out. Laxman dashed over to his side and looked at the screen. The following sentences had appeared.

Summary of conclusions:

- Examined all maps in all categories.
- Verified that four colours are enough in each case.
- Can print out all details.
- Your slow printer will take twelve hours nineteen minutes to print everything.
- Total time of executing this programme: 59.52 seconds.
- End.

The computer had far surpassed all the expectations of Arul and Laxman. When they were completely satisfied with the results they informed Delhi.

The Container Committee duly met in Delhi and studied the report. Most members felt that this extraordinary computer provided unique export opportunities and that the country could and should now go in for manufacture of the prototype. Most, that is, all except Arul and Laxman! They were not interested in the manufacturing aspect. To them the prototype was just the beginning. It was an invitation to study and evolve artificial intelligence, culminating in the von Neumann machine. But they could not express these views openly as the idea of the machine was known only to them and to Professor Kirtikar.

It was Kirtikar who recognized their difficulty and suggested a way out. He proposed that they be allowed to use the prototype for their research for one year while the government set in motion all the legislation and the infrastructure for manufacture. After a year Dr Laxmanan would supervise the production, for he alone knew all its intricate details.

As the chairman was about to conclude the meeting, Navin spoke up.

'Sir! We have to do one important job still. This new baby deserves to have a name.' Navin's suggestion was seconded by others.

'My own inclination is to give it a name of Indian origin, reflecting our tradition, our culture', the chairman said.

Several names were suggested and discussed. Finally, it was Probir Ganguly whose suggestion hit the mark when he said: 'This computer will teach us a lot; so why not call it "Guru"? A name that is simple, yet reflecting our respect for the teacher.'

'Done', said the chairman, noticing the favourable reaction all round. 'May this Guru make us wiser.'

'Amen', said Kirtikar to Arul who nodded in agreement.

As the meeting concluded and Navin stepped out, there was a pat on his shoulder. It was Major Samant.

'A friend of mine is anxious to meet you, Dr Navin. He won't take no for an answer.' He pointed to a waiting car.

Without a word Navin followed Major Samant.

6 The Rendezvous

The Guru of Gauribidnur.

Yesterday a high level committee of scientists, technocrats and administrators decided in New Delhi to go ahead with the fabrication of a photonic computer, it is reliably learnt. The manual for building this super-supercomputer was reported to have been deciphered from the data found in the excavated container at Gauribidnur. Experts rate this computer generations ahead of anything available in the advanced countries of the West. The Prime Minister, who is taking a personal interest in the Gauribidnur project, is believed to have given top priority to the completion of the computer and all being well the first model may see the light of day within a year. However, this crash programme being top secret, no one can be found to brief the press on it. Both the Electronics as well as Science and Technology Departments have kept mum on the project.

This UNI release was, however, sufficient to generate excitement amongst the knowledgeable the world over. Karl Shulz alias Joseph read it in Zurich, while Chushiro Yamamoto saw it in California as he was flying in his private plane.

'Joseph?' Yamamoto was never one to lose a moment. He had called the secret number given by Joseph.

'Who is it?' was the cautious counter-question.

'Chushiro.' This was the pass word agreed between them. The voice at the other end relaxed significantly.

'Joseph here.'

'Have you read the news?'

'I have … don't worry, Chushiro. Everything is under control. It won't take long.'

'Long is a relative measure of time', Yamamoto spoke in his even tone. 'I need absolute estimates, especially where time is money … In your case delay reduces the value of the information, I need hardly add.'

'OK. Message registered. One month', Joseph replied.

'That's too long.'

'But considerably shorter than a year … And with your multinational advantages you can easily make up for a month's lead.'

'One month, Joseph, and no more.'

As Yamamoto hung up he pondered. Yes. The red tape and slow style of functioning of Indian projects would ensure that his one month of delay was of no consequence. As to the computer's capability, Joseph had certainly provided ample proof.

At the other end, Joseph picked up the letter just delivered to him by one of the international couriers. He reread the most important part:

'...Navin has managed to collect most of the manual instructions and I will get a package to you within a week. It will contain everything except the data on the Central Processing Unit. The information on the CPU is heavily guarded ... only Laxman knows it ... but Navin is optimistic.'

Joseph cursed softly at the letter writer. If Pyarelal were asked to stage Hamlet, he would arrange for everything except the Prince of Denmark.

On his return from Delhi, Laxman took up the challenge of making the von Neumann machine. The clues were all given in a highly intricate programme which could not be understood without Guru's help. Not even Arul could get to first base in understanding what was involved. So Laxman struggled along all on his own.

'What research keeps you busy, Laxman, now that you have delivered the Guru?' Navin tried to prod him in the canteen on one occasion as they were having lunch.

'Finding a Guru is not enough, Navin. One must learn from him', Laxman answered guardedly. No one, least of all Navin, must know what he was up to.

'All computers only carry out orders. Yours is no exception. Fast, yes, efficient, certainly. But intelligent? No sir! I think we chose a wrong name for it when we decided to call it Guru. It cannot teach us anything.'

'Wait till you see what this Guru delivers', Laxman suddenly burst out. Exactly what Navin wanted.

'Delivers? Delivers what? May we mortals know what supreme knowledge your Guru will eventually deliver?' he asked in provocative tones.

But he was disappointed. Long before Laxman could answer back, Arul, sitting two places away, suddenly interrupted with a reminder.

'No more discussion, Laxman ... unless you are willing to forego the cricket match ... Come, it is high time we were on our way', Arul almost unceremoniously bundled Laxman away.

'I had to act, Laxman', he explained as they were speeding along towards Bangalore. 'You almost walked into his trap. Remember ... Navin is a bad egg.'

'I got carried away ... I admit', Laxman confessed. But he added, 'I feel strongly defensive when Guru is under criticism. Of course, I should have remembered that Navin is trying hard to get the information about the CPU. Why do you laugh?'

Arul explained. He was reminded of ogres in fairy tales whose life was safely tucked away in some innocent object like a parrot or a fruit. Ogres, who could not be killed in the conventional way. So the hero had to find the secret of an ogre's life.

'The analogy is not quite apt … is it? Here we have Navin the villain, not a hero, trying to find out what makes Guru tick. But joking apart, don't you think your plan of making a robot cannot be kept secret for long? We will soon need technicians to carry out Guru's instructions.' Laxman posed a practical problem.

Arul mused for a while. Laxman's point was well taken.

'The correct plan of action, Laxman, lies in our disclosing the truth but not the whole truth. From computers to automation is a natural course of action. Why not disclose that you are making a robot, a robot of the ordinary kind? But, of course, only a handful need know its real nature.'

'We will raise the matter with the Major today', Laxman said.

'Major Samant? Are we meeting him today? Where?' Arul asked, greatly surprised. The Major seemed hard to avoid.

'Sorry, Arul, I forgot to tell you', Laxman sounded a little embarrassed. 'It was the Major who got us tickets for the one-day international. I like cricket of course, but not to the extent of spending eight hours watching it. As it happens, the match is basically an excuse for us to meet a friend of the Major … from the C.B.I.'

'Who is this man?'

'A Mr Kamala Prasad Singh. He is to meet us in the stadium.'

The Karnataka State Cricket Stadium was overflowing with cricket fans. It was the final and deciding one-day match in the five-match series between India and Pakistan. India had won the events in Delhi and Calcutta, while Pakistan had proved superior in Bombay and Hyderabad.

The two scientists almost turned back when they saw the crowds at the gates. But then they discovered that they had special tickets which entitled them to seats in the VIP enclosure. The gate there was well guarded and they could get in without difficulty. Samant was nowhere to be seen and Laxman had half a mind to go out and look for him. But Arul, who had now caught the match fever, refused to budge.

The game started exactly at 4 p.m. It was to go on till midnight under floodlights. Pakistan had the better luck with the toss and put India in to bat. Judging by the state of the pitch and the Pakistan batting line up, experts wanted India to score runs at an average rate of five per over to win.

The game started on a sensational note. The opening batsmen threw caution to the winds and lambasted the bowling to register twenty-two runs in the first two overs. At this rate, argued the optimists, even three hundred in forty-five overs was not impossible. 'Wait', said the pessimists; for sooner or later this lack of caution will invite disaster.

It did. The Pakistan bowlers discovered their line and rhythm and before long the scoreboard showed 32 for four wickets in six overs. 'Will the Indians even make it to three figures?' wailed a commentator.

'It all reflects on our lack of match temperament', someone behind Laxman spoke in Hindi.

Laxman turned round to find that a tall, lean person had occupied the seat behind him. Laxman did not remember seeing him there when they had come in. He had probably come in late.

As Laxman turned back to watch the match, the man handed him a card. It gave the name only as Kamala Prasad Singh. Beneath it 'PTO' was written in hand. Laxman idly turned the card. There was scribbled message: 'Follow me at the drinks interval.'

7 The Challenge

Arul was deeply engrossed in what he was reading. John von Neumann's thesis in 'Theory of self-reproducing automata' had taken hold of him. Extensive commentaries by experts accompanied the original work. He had first heard of this famous mathematician at school. His maths teacher had set the class a tricky problem: Two railway stations are 120 miles apart. Two trains start at the same time from the two stations and move towards each other along the same tracks. One has the speed of 20 miles per hour while the other moves at 40 m.p.h. A fly that was sitting on the engine of the first train travels at the speed of 50 m.p.h. until it lands on the engine of the second train, whence it rebounds and flies back towards the first train. It keeps shuttling between the trains at the same speed until the engines collide and the fly is crushed to death. How many miles did the fly traverse during its entire journey?

Arul had solved that problem with relative ease. He figured out that the distance between the two trains was steadily diminishing at the rate of $20 + 40 = 60$ m.p.h. For them to collide, this distance (which was originally 120 miles) must reduce to zero. This would take two hours to happen. In those two hours the fly would have travelled altogether $2 \times 50 = 100$ miles.

'Well done!' the teacher had said. 'Do you know how von Neumann, the famous mathematician solved this problem?' The teacher related the tale of how a friend posed the problem to von Neumann, who had almost instantly given the right answer. 'How did you solve it?' the friend asked. 'Why, there is only one method that I know of—the direct one', von Neumann elaborated.

'You calculate all the distances the fly covered from engine to engine. She made infinitely many trips to and fro, but the distance in each trip is reduced in a geometric progression. I summed the series and gave you the answer.'

This direct method requires considerably more calculation, and a reasonably bright mathematician can do the sum in about ten to fifteen minutes. That von Neumann got the answer so fast was typical of how rapidly his brain functioned. To Arul it was more revealing in the sense that his method was the one a computer would use to solve the problem. The direct method—it may involve more calculations—but time was no problem to a fast computer. So von Neumann's brain was ideally suited to think about how a computer would work. No wonder he had proposed the notion of this fantastic machine.

As von Neumann found, there is a basic difficulty in fabricating an intricate machine. For it demands a complicated system of logic which needs a highly elaborate set of cross connections for transmitting information. And the more complicated the logic, the easier it is to make mistakes. From hand-operated machines to automatic ones, the complexity and the likelihood of breakdown both increase. A self-reproducing automaton is of necessity far more complex than an ordinary one.

Realizing this difficulty and the near impossibility of achieving foolproof logic, von Neumann introduced what he called 'probabilistic logic'. It was a system of logic applied to an intricate device, not all the components of which were expected to function correctly at all times. The probability of breakdown of some logical step was thus allowed for.

Based on such a system of logic, von Neumann had proved many mathematical theorems. Some of his conclusions went beyond abstract notions, for he offered explicit constructions for realizing them in practice. A machine with considerable complexity of automation may be able to make another with lesser complexity. But can it make a copy of itself? 'Yes', said von Neumann, and proved it mathematically. His construction of such an automaton was, however, too complicated to be within the scope of current technology. At least so the experts thought, and Arul agreed with them.

But what about future, more advanced technologies? If they could make a self-reproducing automaton, what problems would they pose? Von Neumann's work raised such questions. Would these machines follow the evolutionary doctrine applicable to living systems, wherein species improve through natural selection and interaction with the environment? Can one generation of automaton 'learn' from the mistakes of the previous generation?

To Arul one question kept coming back again and again. If the automata strove to improve themselves at all costs, how would they deal with anything that interfered with that goal—even assuming that the interfering agency was human?

'Laxman, at your suggestion, I studied von Neumann's thesis. I now appreciate the enormity of the problem you are trying to solve. Without the blessings of the Guru, I would say that your task is an impossible one', Arul said.

'For me, the main work has already been done by those who left the container behind. I am merely trying to understand their prescription. Even that is proving hard enough', Laxman replied.

'Ha! Ha! It's very amusing to see you being reduced to a computer that follows programmed instructions.'

'Not even that! The computer at least understands the instructions. I have yet to reach that stage. Which is where Guru helps me.' And Laxman pointed to a locker.

The sanctum sanctorum of the computer room was accessible only to Arul and Laxman. Strict security, manual and automatic, ensured that nobody else could enter. Even in such a guarded place Laxman had a specially made locker.

'Over there, further instructions are to be fed to the robot I am making. According to Guru, the information has to be fed in a certain order to be strictly adhered to.'

'That is fine so far as it goes! But Laxman, I have a word of caution. You made Guru as per the instructions. So far so good. We know what computers are. This one is highly exceptional, but still not totally unfamiliar to us. But what you are making now is new ... alien.'

'Bah! You are probably reading too much sci-fi, Arul. Alien, indeed! This robot is after all a robot, an automaton. One that can be controlled from without—not Frankenstein's monster.' Laxman laughed.

'Nevertheless, I must advise caution. Don't get carried away.' Arul's face expressed genuine worry.

'OK! Let me promise to consult you at each important step in the project. How is that?' Laxman asked, still in a light vein.

'Good, but not good enough. We need an experienced mind. I would greatly value views from Professor Kirtikar. We consulted him in the beginning, remember?'

'Done, Arul, done. To hear you speak, I am beginning to feel as though I am making a nuclear bomb instead of Vaman.'

'Vaman?' Arul was hearing the name for the first time.

'My projected robot. I have already named him Vaman—for the instructions as interpreted by Guru make him only about a metre tall. What do you think of the name?'

Vaman! One of the ten incarnations of Vishnu when he appeared as the Dwarf God.

'Appropriate ... very appropriate', said Arul thoughtfully. The name had other connotations that worried him momentarily. According to the Hindu legend, King Bali offered Vaman as much space as he could cover in three steps. Thereupon Vaman grew, and grew, so that his three steps covered the heavens and all of Earth, with the result that Bali was deprived of his entire kingdom and had to retreat to the underworld. Would Laxman's Vaman develop sinister intentions? Arul had vague misgivings, but he dismissed them as he nodded heartily and said:

'Three cheers for Vaman!'

Vaman

1 The Monads

Laxman thought that even Guru would not be able to cope with the complexity of instructions needed for the making of Vaman as, once again, he encountered the verdict 'Programme Error' on his terminal. But from past experience he had learnt to respect Guru. Previous programme errors had turned out to be traceable to his own mistaken execution of Guru's instructions. On each occasion Guru had correctly interpreted the Vaman code, as Laxman called it.

The complexity had increased as he progressed. There were occasions when he felt like giving up, but the tantalizing goal egged him on. That he was making a robot was now known all over the Science Centre. Like Guru, the robot was expected to be exceptional. But nobody knew in what way—certainly no one even imagined that a thinking, self-replicating robot was in the making.

'How long before we may expect the incarnation of Vaman?' asked Navin on one of those frustrating days.

'A couple of weeks, at least. The hardware problems of his brain still need to be sorted out', Laxman answered.

'But how about your own work, Navin?' Arul asked, knowing full well that this was the only way of diverting Navin from his questions.

He was right. Navin was at heart a research worker. He relished narrating how he had deciphered archaeological clues to reconstruct an era that was long past. As Navin embarked on his elaborations, Laxman broke in: 'Navin! You have so much to say. Why limit it to canteen table talk? It is about time you addressed the entire staff of the Science Centre on all your finds.'

'Absolutely!' said Navin. 'I am very keen to give a detailed talk so that I can share my excitement with others. I will do so once all my slides are made.' And he rushed off on his errand.

'Navin has changed', Arul commented as his figure disappeared through the door.

'Yes. The Major's medicine has worked. He has decided to play ball. He has been dutifully conveying my false data on Guru's CPU to Pyarelal. The Major threatened him with god knows what, but he has turned the corner', Laxman replied.

'I wonder what our friend Yamamoto will say once he realizes that he has been duped', Arul mused.

'It is not as simple as that, Arul', Laxman smiled, for he was about to reveal his secret. 'The information is so complex that it will take Yamamoto and his bunch of experts a considerable period to realize that it is basically defective. They will continue thinking that their implementation is at fault. And meanwhile, they have no option but to rely on what we supply them with. Indeed, Samant is hopeful that once Vaman's reputation gets about, Shulz will pay us a visit ... and walk right into a well-laid trap.'

'But what is to prevent Shulz from operating from afar—he need not enter India. He has that agent of his, Pyarelal', Arul retorted.

'Samant does not think so, and I agree with him', Laxman said. 'Vaman will be a prize that Shulz cannot entrust to others—it is too important to him for that. It will be the supreme achievement of his nefarious career. Vaman will be the bait for Shulz.'

'I hope you're right, Laxman. I hope, too, that Navin has really turned over a new leaf. It is a dangerous game that Samant is playing', Arul said.

'May be you are right. In any case it is necessary for both of us to be very cautious. But as far as strategy is concerned, I leave it to the experts. Samant evidently knows what he is up to. Well, I must be on my way back to the drawing board, Arul.'

Laxman got up, but Arul remained seated, a thoughtful expression on his face.

The auditorium of the Science Center was packed to capacity. This was the occasion when the staff was to be briefed about the contents of the box which had started all the activity. Exactly at 5 p.m., Navin rose to give his presentation. He had come armed with slides and viewgraphs.

'Friends,' he began in a voice scarcely concealing his excitement, 'this container at Gauribidnur is without doubt the greatest archaeological find ever. Of course, you are dazzled by the supercomputer Guru and looking forward to the arrival of Vaman. The technological fallout of the container is indeed fantastic, but to a person like me the past holds all the allure. So let me take you back to the ancient times when the people who left the container behind lived and flourished ...

'I was intrigued by them right from the day I set my eyes on the container. I will not bore you with technicalities, about the methods commonly used by archaeologists to interpret the relics they find. Instead I will give you the final outcome of my investigations in ready-made form. Those who want to know the why and wherefore of it may reserve their questions to the end.

'I estimate that this civilization is about twenty thousand years old. All our so-called ancient civilizations—of Egypt, Babylon, China, or Harappa-Mohenjodaro—are modern compared to this one. People argue about the exact time of the Vedas. But even with the oldest quoted estimate, the Vedic era occurred much later than this civilization.

'These people called themselves Monads, after Mona, the Earth in their language. It was a worldwide culture, transcending regional differences of geography. There were no nation-states, nor were there any tensions or quarrels between different regions. It was truly one family of people inhabiting our planet.

'There are two ways of estimating how advanced a civilization is. One method uses the measure of energy consumed by a civilization, not only to maintain its standard of living but also in the exchange and broadcast of information. On this scale, the Monads were a thousand times more advanced than we are now. For comparison, the gap between the Greek civilization of two millennia ago and us is about the same.

'The second method estimates the total information content of the civilization in question. Knowledge *per se* can of course defy objective evaluation, but the language in which it is expressed lends itself to quantification. For those unfamiliar with computers and binary arithmetic, I will digress a little to explain this point.'

Navin paused to sip some water and then projected his first transparency on the screen. It showed positive (+) and negative (−) signs distributed in thirty-two rows, with five signs in each row:

$$
\begin{array}{ccccc}
+ & + & + & + & + \\
+ & + & + & + & - \\
+ & + & + & - & + \\
+ & + & - & + & + \\
+ & - & + & + & + \\
- & + & + & + & +
\end{array}
$$

He then continued.

'On the screen you see thirty-two different alternatives for distributing pluses and minuses in five places. Why thirty-two? The answer is simple. Each of the five places has two alternatives, either a plus or a minus. Since each alternative for the first place can be combined with each alternative for the second and so on, there are in all $2 \times 2 \times 2 \times 2 \times 2 = 32$ ways of arranging these signs.

'What has all this to do with information, you may ask. Well, in a computer's binary logic, the basic information consists of the alternatives 'yes' or 'no' to each question. These are my plus and minus signs. Each set of alternatives is called an information bit. What you see on the screen are the thirty-two different sets of alternatives of information available through five bits.

'Now we can identify each of the quintets with a letter in the English alphabet together with six signs like the full stop, comma, etc. A four-letter word will thus need twenty information bits, and a ten-word sentence of forty letters, two hundred bits. A book of fifty thousand words will need approximately one million bits.

'Let us estimate the total number of different books in the English language as around ten million, based on the inventories of the biggest libraries in the world. Their information content is around ten million million bits. Taking into account books in other languages and also the fact that other modes like pictures and music also carry information, the total information content of our civilization is no more than a million billion bits.

'Again, this is about a thousand times more than what the Greeks had. The Monads, on the other hand, were a thousand times better informed than us!

'However, their numbers were limited—no more than around a hundred million. Therein lay the cause of their prosperity. They could and did control their population and used the resources of our planet judiciously. Their colonies were typically of twenty to thirty thousand people. Their main energy sources were two: the direct exploitation of solar energy and the fusion reactors that our scientists are striving to construct today.

'With this introduction I now show you the slides I got specially made based on the information supplied by the container. I will also give some factual information on the transparencies.'

So for the next hour Navin held forth brilliantly. His talk was followed by numerous questions, the last of which came from Arul: 'Why did they bury this time capsule in Gauribidnur, of all places?'

'As far as I can make out, this place was an important administrative centre for the Monads, one of the ten dotted all over the globe. They wanted to choose a site near an administrative centre and fulfilling certain conditions. The place had to be free from earthquakes, as this one is. The rock strata here are remarkably stable. So anything buried here would remain undisturbed. The Japanese location, for example, did not meet this criterion. Next, they wanted the underground water table to be very low—which ruled out their European centre near Holland. A third criterion was that the container should be well away from the seashore—satisfied here, but not by their centre in what we today call Florida ... Well, the long and the short of it is that, after these and many other considerations, our spot right here turned out to be the best.'

'Ah, that explains it', Arul replied after Navin had finished.

'Explains what?' Navin asked.

'Why did we discover the container here? My criteria for housing the gravity experiment were precisely these! On the basis of these I chose Gauribidnur as the best site. So our digging here and finding the container was not as great a coincidence as I had imagined', Arul said.

After Navin's talk, Arul and Laxman went over to the latter's living quarters. Urmila joined them shortly afterwards.

'And where have you been my little maid?' Laxman asked.

'Why, to the lecture, of course. Wasn't it magnificent? Now I know why Navinji is so much in demand as a speaker. He can feel the pulse of the audience—what a contrast from my dear hubby', Urmila answered, squeezing Laxman's hand.

'Laxman, it looks as if your talents are not appreciated at home', Arul added jocularly.

'I gave up lecturing to her long ago. She can't understand even the simplest things on computers', Laxman said.

'Arul, you should ask your friend what he means by the "simplest". His idea is to quote some obscure algorithm to start with and follow it up with totally incomprehensible questions', Urmila put her case.

'Well Madam, here is a simple question which even you can comprehend: when are we to get our coffee?'

'I am like your computer, sir—I can execute orders only after you have given them. So you will have to wait fifteen minutes.'

As Urmila went into the kitchen, Laxman put on an LP of Beethoven's fifth symphony. Urmila's return with the coffee brought the two scientists out of the respective reveries which the profound music had plunged them into.

'Arul, perhaps you can answer my question in simpler language than my husband can', Urmila asked, as they sipped the delicious 'real' coffee that no 'instant' version can ever emulate.

'Go ahead, I will try', Arul replied.

'Navinji did not make one point clear … Why did the Monadic civilization, which was so advanced, become extinct?'

Before Arul could reply Laxman broke in: 'Umi, your question is basically illogical. Navin's talk was based on what he found in the container. Evidently, whatever was in the container was left by the Monads when they were alive and flourishing. How can you expect to know from that how they died? It is as ludicrous as expecting a person to write his own obituary.'

'Arul, here you see an example of Laxman's obscurantism.'

Arul, however, was thoughtful as he replied slowly. 'Urmila, Laxman's technical objection apart, your question is basically a valid one, and worrying at that. In the accounts found in the container we see a picture of a well run society. Evidently the Monads had to face some unexpected natural catastrophe. What could it have been? An ice age? A major earthquake? Or a volcanic eruption? Did a meteorite or a comet hit Earth? It had to be something that could not be overcome by the Monads with all their very advanced technology and meticulous planning.'

Laxman shook his head in disagreement. 'Talk sense, Arul. Natural disasters can be catastrophic, I admit that. But a really big event like that is bound to leave a mark or two behind. Moreover, twenty thousand years is not really long enough to obliterate those marks. Why don't we see any relics today?'

Laxman's point was well taken. Arul had no answer. Why were the Monads wiped off the face of the Earth? The container, as Laxman had argued, would not be expected to provide the answer. How and where would they find it?

2 The Little One

'Let us see how your little prodigy performs.'

The words were Navin's, uttered in his typical half jesting tones. But like Arul, he too felt the excitement of the occasion. It was Vaman's first trial performance.

The metre-long robot was lying flat on the work bench. Not far from him was a computer terminal for giving instructions, instructions that were to be conveyed to Vaman by microwaves. His brain had a receiver that could receive instructions and respond suitably.

'Vaman, open your eyes.' Laxman's instruction appeared on the screen.

'Bravo', shouted Arul, greatly excited and pointing to the bench. Vaman had opened his eyes. He was staring straight up at the ceiling. 'The little fellow responds even in his sleep', muttered Navin, equally impressed.

'Sit up', was Laxman's next instruction on the terminal. Slowly, Vaman performed an almost human action to heave himself into sitting posture.

'My next instruction is more elaborate', said Laxman as he typed furiously on the computer: 'Slide to the north end of the table but take care not to fall off.'

Vaman began to slide to the correct end. He evidently had a built-in direction finder. But what was more important, he went so far that his centre of gravity was just within the edge of the table—the farthest he could go without falling off.

'Get down', Laxman winked as he typed the next instruction.

Vaman slid forward and fell clumsily to the ground. Laxman picked him up and put him back in his original sitting posture. He then issued the same instruction. How would Vaman respond now?

'Fantastic!' Arul really was impressed as he watched Vaman carefully get down and stand on the floor. 'That is what I wanted to check! The fellow learns from his mistakes', Laxman added by way of explanation to Navin who looked puzzled.

'Let me play chess with him, then' Arul said. In his student days he had reached the level of a 'Master' and would have scaled greater heights had he devoted more time to the game.

'Surely. But in due course, Arul. We will explain to him the basic rules of the game, outline a few stratgies, brief him with a few famous games played by experts. Then we will let him loose on you', Laxman said with a grin.

Laxman's unstated expectations were realized. Having learnt the basic rules of chess and studied the styles of such great players as Alekhine, Capablanca, Fischer, and Karpov, Vaman rapidly improved his competence. He could learn from his mistakes and plan strategies several moves ahead. He lost or drew with Arul in their first ten encounters. As it turned out, he needed this experience to gauge his opponent's competence. Having placed Arul on his own scale, Vaman adjusted his game so that he scored a victory in their eleventh game. And from then on he never looked back.

Vaman also acquired increasing flexibility in operating his limbs. The commands issued by his 'brain' became more precise and his body responded accordingly.

The next major step to be followed according to Guru's instructions was to train Vaman to respond to verbal orders from Laxman. The sound waves were converted to microwaves by Vaman's 'brain' and interpreted for action. Thanks to the precise details followed in his make-up, even this complicated step worked out well. To identify Laxman's voice, Vaman needed a password which Laxman would whisper before beginning his orders. No one else could give orders to Vaman.

It was not long before Vaman's abilities became well known all over the Science Centre. From carpentry, to repairing TV sets or other electronic equipment, to gastronomical feats in the kitchen, Vaman picked up everything he was taught. Even sewing and darning were part of his repertoire. Nevertheless, another step remained before Vaman could become a von Neumann machine. He had to be taught to think for himself, to make his own decisions. Laxman was impatiently waiting to use one of the top secret packages in his possession to reach that stage. But, as he had promised Arul, he would have to consult Professor Kirtikar before taking such an important step.

His plan to have that crucial discussion with Arul and the Professor had to be postponed, however, thanks to an unforeseen interruption.

The interruption, though unexpected, would have come sooner or later anyway.

It all started with unofficial news reports. As Vaman's achievements were now well known in the Science Centre, it was inevitable that accounts about him should leak to the national press—garbled accounts that led to public debate, culminating in questions in parliament. What was Vaman? How was his origin related to the archaeological finds at Gauribidnur? Who was controlling him? Would the P.M. enlighten the House on these crucial matters? The clamour for an answer transcended party lines and the Speaker had no option but to admit the questions officially.

The P.M.'s secretariat got busy. Probir Ganguly, Harisharan and Raj Nath, the three concerned relevant secretaries, were asked to furnish details. The three met, deliberated and finally decided that one of them should visit the Science Centre and collect the information first hand. The task fell on Raj Nath.

'I'm glad you were chosen, Dr Raj Nath. I could not have stomached the silly questions Harisharanji would have asked', Laxman said as they drove together to Gauribidnur.

Raj Nath lit his pipe and brooded before making his reply. 'Not that my questions are going to be any wiser. But Dr Laxmanan, as a once-practising scientist I can understand your feelings. A serious researcher gets upset when interruptions like this occur, more so when bureaucrats are sent by politicians to question him. But do take a wider view. Who funds your research? The nation, whose elected representatives vote the money. The bureaucrats merely provide the mechanism for a two-way dialogue between the funder and the fundee.'

'Agreed, Dr Raj Nath. In a democratic system this is the procedure … one can't avoid it. But there are bureaucrats and bureaucrats. Why are Harisharans the norm and you or Probirda the exceptions?'

'Ah …' Raj Nath let out a smoke ring by way of an answer and asked, 'Do you have anything better to suggest?'

Laxman didn't. That a person who is given public funds should be answerable to the public was perfectly reasonable. But how can one establish a sensible procedure for dealing with sensitive scientists like himself? Scientists who strove to achieve excellence without compromise?

Seeing that Laxman was silent, Raj Nath continued: 'Look at it this way Dr Laxmanan. In this project you have got all the facilities because the P.M. backed you all the way, overriding bureaucrats like myself. Why? Because he appreciated the significance of the relics pretty early on. Don't you think his confidence in you needs to be reciprocated?'

'Absolutely! In fact, I would suggest the P.M. visits the Science Centre. All of us, Vaman included, will be honoured by that.' Laxman responded spontaneously.

'Well spoken, Dr Laxmanan. Let us work on the idea', Raj Nath said heartily. His mind was already working out the various arrangements needed for the P.M.'s visit.

'O.K. Chushiro, I promise.' Shulz was obviously agitated, for he put down the receiver with a bang.

In front of him was a Reuters news flash—probably the same one that Yamamoto must have seen. 'The fellow does not waste time, does he?' he muttered to himself as he reread the item:

The Prime Minister of India today visited the Science Centre at Gauribidnur. Vaman, the newly made robot, was introduced to him. The P.M. congratulated the staff of the Science Centre, especially Dr Navin Pande, Dr Arul, and Dr Laxmanan, for their efficient evaluation of the relic container and the follow-up action which resulted first in the photonic computer Guru and then the robot Vaman. The P.M. expressed confidence that the information from the container would launch India on the way to becoming a top class technological nation.

Yamamoto was right to be peeved. His version of the supercomputer was still not working. Pyarelal had not delivered what he had promised—in spite of repeated warnings. Instead, he had supplied misleading information ....

Pyarelal must be dispensed with. Of course, the real culprit was Navin. For some reason Navin had let them down. All this disinformation was being supplied by him ... Given the choice Shulz would have eliminated Navin first. But he had no choice. Navin was inside—in the vital place, to be effective if he chose. If he saw what happened to Pyarelal, he could be persuaded to change his mind, thought Shulz grimly.

Shulz rapidly thought out his plan of action. He resisted the temptation to visit India and be there on the spot. That must wait until Pyarelal was taken care of.

From his highly select little telephone book, Shulz picked a Bombay number ... An underworld contact who owed him a return present.

3 The Precaution

In the presence of Professor Kirtikar and Arul, Laxman opened the sealed package. The next stage in Vaman's evolution was outlined here.

'The information appears to be in two parts' Laxman said, while Arul busily started translating the details from the ancient script. Thanks to Guru's training, both Arul and Laxman had gained considerable competence in understanding that language. Arul was soon at a terminal.

'The first part enables Vaman to think for himself while the second teaches him how to make his own copy. Of course, for the second part we do nothing but feed him with instructions. He could figure them out himself ...' Arul was reading the information that had appeared on the terminal.

'So part two must necessarily come after part one', Kirtikar said.

Even the first part leading to a thinking robot contained information that was several decades ahead of the present knowledge on artificial intelligence, thought Laxman. Like Arul, Laxman was feeling that they were once again at a step that marked a quantum jump in mankind's ascent of the technological ladder. They both looked expectantly at Kirtikar, who had begun to walk up and down that small office.

'Which scientist can resist a fantastic discovery especially when it is presented on a platter like this?' mused Kirtikar, trying to keep pace with the rush of conflicting thoughts. 'I can understand your excitement. But two decades separate us. Call it a generation gap if you like! And I cannot help feeling uneasy ... are we going to repeat the episode of Bhasmasur from our Puranas?'

Bhasmasur ... the monster created by Lord Shiva, a monster who could turn to ashes anything he put his hands on. And eventually Bhasmasur threatened Shiva himself.

'I too thought of the same story, Professor Kirtikar', Laxman smiled as he said this. 'But Shiva was, after all a very naïve god who could be easily taken in. We can be more cautious. Vaman, whatever he does, needs energy to do it. He derives it from his in-built battery which can be charged—and is always being charged—by any kind of light. But any mechanical device, however sophisticated it might be, can also be rendered ineffective. I can arrange to render his batteries useless by a remote sensing microwave device.'

Professor Kirtikar looked at Arul, obviously for a second opinion. 'Yes. That is possible. Before we feed Vaman with the first part of the package, we should try out such a device and implant the appropriate receiver in him.'

Kirtikar passed a hand across his brow. It was clear that both the younger men were keen to go ahead: Laxman more so, because this was a dramatic step in his field. That Arul also seemed keen could not be ignored. Although he possessed it, Kirtikar could not bring himself to exercise the veto. True, Laxman and Arul had come up with a fairly satisfactory preventive device in the unlikely event of Vaman going out of control. But would it really work when needed?

'Well, I have to go along with you two after all', he said, finally making up his mind.

'Shake hands!' Laxman broke out of his usual reserve in his presence ... excitement had taken full hold of him.

Keeping the first part on the table, Laxman carefully repacked and sealed the second and restored it to safe custody. He then went over with the first part to his lab.

'Arul, what do you honestly think? Will Laxman's safety mechanism work when it comes to the crunch?' Kirtikar asked Arul.

'When? You mean "if" don't you? Arul was startled by this query. 'A device using remote microwave control is very straightforward. Why should it fail? If it were something complicated ... of course. One has to allow for failures ... But this one? Surely not!'

However, even as he said so, Arul became less sure. Kirtikar rarely raised technical objections; but whenever he did, he had almost always been justified. Even now he voiced his reservations. 'The very simplicty of your gadget may mean that it could be easily spotted by Vaman. After all, a thinking robot is almost human. Like us, he will want to protect himself. When he discovers that he is harbouring a debilitating device ...'

'We will have to ensure that he does not discover it so easily, Professor Kirtikar. But your point is well taken', Arul said thoughtfully.

'I think we need to go further, Arul ... and that, too, without informing Laxman.

'Why?' Arul asked, shocked. With Laxman he had had no secrets so far.

'Because Laxman looks upon Vaman as his creation. Unconsciously, he is emotionally involved with him. Even as a scientist, his sole criterion at present is to make his experiment a great success ... to make Vaman as efficient as possible. It's not his fault. All dedicated scientists are like that.'

Still, Arul was unsure. A clear, logical mind like Laxman's ... would it succumb to ego, to the extent of not foreseeing the catastrophe should Vaman turn belligerent?

'Remember the Mahabharata', Kirtikar sought to drive home his point. 'At Duryodhan's birth there were all kinds of ill omens and his parents were advised to get rid of the child as he would otherwise cause everyone great misfortune ...'

'Come, Professor Kirtikar! Surely you don't believe in omens and astrology?' Arul said.

'I don't. But that is not the point. Dhritarashtra and Gandhari, the boy's parents, did. But they still could not bring themselves to abandon their first born. Vaman is Laxman's first born, after all! What I have to suggest now had better be kept secret from him.' Kirtikar was totally serious as he said this.

At last Arul caught his drift and was again shocked. You are not suggesting that we …'

'I am … if the need arises, there is no alternative! Let us hope the need does not arise.'

Urmila was watching a Tamil film on the VCR. The video recorder was one of the few concessions Laxman had made to alleviate the monotony of her existence in the Science Centre. For a couple of weeks after its arrival the equipment served its purpose as Urmila watched one film after another almost continuously. But eventually the films brought their own monotony of essentially the same set plots with minor variations. Today she had decided to lock the equipment away after the film was over. No more would she ask Laxman to send someone over to Bangalore and get a few more tapes from the video libraries there. Enough was enough!

'Umi … here is some fun for you.' The unexpected voice of her husband behind her made Urmila get up with a start. Yes. She was not dreaming. Laxman was there in person. She was even more surprised to see that he was not alone.

'I brought Vaman along since he wanted to meet you.' Laxman indicated his little companion. It was a break with routine in more than one way. For the first time Laxman had dared bring Vaman outside the main office building.

Urmila did not quite know how to greet a robot. It was Vaman who took the initiative. He joined the palms of his hands in traditional fashion and said, 'So delighted to meet you, Urmila. You are a lucky man, Laxman.'

'Why … when did you start talking, Vaman?' Urmila burst out of her reserve. Vaman's voice was somewhat mechanical, with hardly any inflexions. Nor were there any emotional overtones in it. But the delivery was perfectly clear.

'Since the day before yesterday. Laxman gave me my personality that day. I have evolved a good deal since then … so much so that he feels that I may now talk to other human beings. Naturally, I chose you as the first non-scientist to talk to.'

Naturally! What self-confidence the statement implied. Yes, Vaman had definitely changed from the efficient but mechanical robot she had seen earlier. He had, as he just said, acquired a personality.

'Well spoken, Vaman!' Laxman said with the beaming approval that a teacher feels towards a student prodigy. Then he felt that he owed an explanation to Urmila. So he added, 'This little guy, Umi, is not just an ordinary robot any more. He can think for himself. And he is a glutton for knowledge. … It is all I can do to pull him away from Guru. He is learning so fast that soon

I will have no more to teach him. Nor will Guru have any more information than what Vaman has stored in his little brain.'

'Sister, if I may call you that … Laxman is exaggerating somewhat', Vaman intoned. 'As an ordinary robot I did only what was ordered, may be more and more efficiently as I repeated the same tasks over and over again. Now I can act on my own, without Laxman's password and order. I am aware of the nature of information fed to me, aware of what it can lead to and what its limitations are. I can use it effectively when needed. Initially this awareness brought confusion. Now I am able to assess and act. I need to know more and Guru supplies me with more.'

'But, Brother Vaman, what is your purpose in seeking so much knowledge?' Urmila asked.

'Why, as I said, to make myself more effective in serving your needs. As Laxman told you, I am beginning to be more efficient than even he could have imagined.' This was said by Vaman in a matter of fact manner, without any trace of self-glorification.

'That you are, sonny boy!' Laxman beamed as he patted Vaman on the head. He then addressed Urmila. 'Let me tell you of his recent encounters. You know Raghavan. How he prides himself as a trouble-shooter when there is any mechanical breakdown. Today, as he was busy trying to locate the fault in our big dynamo, this fellow came along and spotted it within minutes, I am going to try him out on our medical officer later today.'

'What, are you letting him loose on Dr Antia?' Everybody on the campus was overawed by the M.O., not least Urmila.

'Why not? It will pull him down a peg or two!' laughed Laxman. Then he became more serious as he added, 'But, Urmila, the best so far is what Vaman did to Arul. He solved in a flash the mathematical equation Arul had been struggling with for over a week. Arul was so impressed that he took out his ring and put it on Vaman's finger, saying that henceforth Vaman was intellectually his superior.'

Vaman showed the middle finger of his right hand where a gold ring sat somewhat incongruously. 'I will always greatly value this present, Urmila. It is not often that we robots get such an honour.'

'You deserve it Vaman!' said Urmila as she looked at the ring admiringly.

She dismissed a passing thought that puzzled here. Like most women, she prided herself on being very observant where human beings were concerned. Yet she could not recall ever seeing the ring on Arul's hand before … but then, she told herself, what with her recent video craze, she had not met Arul for several days.

4 The Decoy

'No, Dr Navin. There is no mistake with this message. Pyarelal is dead ... murdered', Major Samant said seriously as he fingered the sheets of paper on his desk. One sheet contained a scrambled message from Kamala Prasad Singh, the other had Samant's unscrambled version.

'But he was always so careful. He had bodyguards ... he had survived several attempts on his life. I still can't believe it.' Navin was deeply shocked.

'You will, when you see press cuttings from the Delhi papers tomorrow ... But they won't add much to what we already know. And certainly they won't suspect the reason, as I do.'

'Reason? Isn't it obvious? Some gang warfare with the Bombaywallahs getting into it?' Navin asked.

'That is for the papers. These rivals don't normally operate like that. They don't infringe on one another's territory. No, there is more to it than meets the eye.'

'My eye, certainly', said Navin.

'You will soon see. The killers were induced by heavy payoffs ... from abroad. From our dear friend Shulz.'

Navin was speechless with amazement. Why should Karl kill Pyarelal who was an important link in the chain to the Science Centre ... to him?

'Pyarelal was killed because he kept feeding false information. The information you supplied, Dr Navin', Major Samant said in even tones.

'I did so at your suggestion', Navin said—half in accusation, half in fear.

'You did. I admit I overplayed the hand. I was expecting Shulz to merely abandon Pyarelal and try to establish direct contact with you. That he has overreacted to the extent of having Pyarelal killed means he is himself under pressure ... from his clients. However, the end result is the same. The decoy is dead ... the tiger will soon come out in the open.'

'What do you mean?' Navin visibly paled as he pictured Karl Shulz after his blood next.

But Major Samant was on the internal phone.

'Yes, Dr Laxmanan. Please come right away. I am also calling Dr Arul ... And don't forget to bring Vaman, too.'

'Vaman? Why Vaman?' asked Navin.

The Major motioned him to silence. He obviously wanted time to think, to react to this new situation.

In a few minutes Arul and Laxman rushed in, followed more ponderously by Vaman. In a few crisp, well chosen sentences Major Samant appraised the first two of what had happened. Then he concluded: 'I asked you to bring Vaman because whatever Shulz may do next will concern him.'

As Arul and Laxman took it all in, Vaman said, 'I'm afraid that I am in the dark about these people. Who is Pyarelal? Who is Shulz? Why do you suspect that one killed the other? And why does it all concern me?'

The Major looked at Laxman. He had always been somewhat at a loss when talking to Vaman, whom he could not accept as an intelligent robotic counterpart of humans. So Laxman had to brief Vaman about the roles of Navin, Pyarelal, Shulz and Yamamoto.

'So Pyarelal lost his life for supplying false information—what Navin had supplied as genuine', Vaman summed it all up.

'Under my instructions', Major Samant said, to make Navin feel less uneasy. 'I take full responsibility.'

'But why are these outsiders like Yamamoto after Guru?' Vaman asked.

'So they can commercialize it. They will make enormous money making and selling it ... And now that they know that Guru helped in making Vaman, their expectations have increased. They can very well see how useful Vaman can be', the Major said.

'This means that they will intensify their efforts to get at the CPU ... the correct CPU of Guru', Arul said.

'And they will also try to get hold of Vaman', the Major added. 'That's why ...'

Vaman broke in suddenly with another question, the suddenness of which surprised everyone present: 'But what is wrong with more Gurus and more Vamans? Can't you humans employ them more fruitfully to your benefit?'

It was Arul who chose to reply. 'Sure we can. But will we? These things can be used rightly or wrongly. Considerable thought must go into how we use Guru, and how we use you, Vaman. We do not want to make any mistakes.'

'So we cannot afford to let you fall into wrong, unscrupulous hands', Laxman added.

'Exactly! And hence I am immediately going to take very strict security measures ... till we catch our tiger. Mr Navin, from now on you are not to establish any contact with Shulz. And all residents of this complex are going to find it very hard to get out. This applies to present company also ... perhaps even more so.' Major Samant rose to indicate that the discussion was over and that action was at hand.

5 The Watcher

Dear Lalitha

For the first time in our correspondence since I came to Gauribidnur, I have to report a break in the dull monotony of my life!

It started a few days back. Rakhee, my maid-cum-cook, approached me just after we had returned from a shopping trip. (As I mentioned before, she is one of the Major's 'spies' who looks after me!)

'Madam, did you know that you are being followed whenever you go into Gauribidnur these days?'

'Of course. You know that I know that both you and Jeevan are my shadows,' I replied teasingly. But Rakhee was serious.

'Not us, Madam! I was referring to others whose intentions I am beginning to doubt', Rakhee said as she readjusted her saree.

My face must have registered sufficient incredulity for her to continue speaking. 'Two incidents occurred, one last week and one today as we were out shopping. Do you recall when we went out to buy ice cream?'

'That was last Wednesday', I recalled.

'Right, Madam. I was following you with a large slab of ice cream in a thermo-cole box …'

'And a person bumped into you and you dropped the box. Fortunately, it was tied up securely with string so that no damage was done to the ice cream. But as I recall, Rakhee, the fault was yours, wasn't it?'

'It was, Madam. But do you recall the person I had bumped into? Fair complexion, grey eyes, dark hair, longish, unkempt … height about five feet six inches', Rakhee said.

'That is for your eyes, trained as you are in police work. For myself, I don't remember all these details', I confessed.

'Today, when we went to get your watch repaired, the same man was standing on the opposite footpath', Rakhee said. I, however, dismissed the matter right away.

'May be, may be not … But I don't think it significant. Both these spots were in the centre of the town. May be the man lives or works in that area.'

There the matter rested. I thought that Rakhee was becoming paranoid. After all, the poor girl is wasting her time watching me here … and nothing happens!

That is, until the day before yesterday! Rakhee and I had been in the outskirts of Gauribidnur, the other end from here. We had gone to a farm where fresh sweet corn was available. As we completed the deal and made our way back to the jeep, I saw him myself. He was standing across the road, chattting to the small stall owner selling betel leaves and cigarettes. Excitedly, I drew Rakhee's attention to him. But somewhat brusquely she bundled me back into the jeep.

'I know, Madam. I saw him the moment we set off for this farm', she said as the jeep took off. 'But on our part we should not let him suspect that we have spotted him.'

'Why?' I asked somewhat stupidly.

'He will be replaced by somebody else and we will have to start all over again. In our profession, Madam, it is better to deal with the devil you know.'

'Do you know this person, Rakhee?' I asked.

'I guess so, but I must make sure first.'

How she proposes to do so, I don't know. But for myself I can no longer dismiss this last incident as coincidence. The farm is not in the centre of the town. The neighbourhood there is not so populated. What was this man doing there?

And today, Lalitha, we have all been served with fresh orders, amounting basically to house arrest! Some other things have happened that I don't know of and the Major is not taking any chances.

May be he will not allow this letter through! But I had to write it to clear my own thoughts. If you do get it, let me add our best wishes to all three of you.

Yours in captivity,
Urmila

As Urmila finished her letter and put it in an envelope, Rakhee rushed into Jeevan's room in the outhouse.

'It is confirmed!' he added calmly. 'You did a good job, Rakhee.'

He had a number of photographs on the table. To untrained eyes they would have looked like portraits of different people. He took out a picture of a long-haired man with grey eyes.

'This camera they gave me in Delhi is superb for telephoto pictures ... You would not believe this was taken from about two hundred feet, would you?' Rakhee said.

'The Major has confirmed that we are dealing with our old friend Balu ...'

'Alias the Nawab, or Chhajjuram, or many other incarnations' added Rakhee, looking at the various pictures.

'He is out on parole after serving a little over half of his twelve year sentence for girl snatching. The fellow has not wasted time since coming out ... But what interests him here?' Jeevan mused.

He decided to leave that to Major Samant to figure out.

'Vaman, can you and will you solve my problem?'

Both Arul and Laxman were surprised by this unexpected interruption from Urmila, who until now was no more than a quiet bystander at their discussion about Vaman's future. Vaman, however, responded in his matter of fact way:

'Your problem, sister? State it and I will do my best.'

'Then take me out of this prison, Vaman. My husband cannot do so. Perhaps you can?'

This was a challenge to Vaman, but, to avoid losing the case by default, Laxman butted in.

'Vaman, she is asking you to do the impossible. You know very well the delicate situation we are all in. Urmila no less than others. After he discovered that the notorious Balu is shadowing her, the Major simply refuses to let her out—even under an escort. I have entreated and threatened, but Samant will not budge.'

Vaman paused a while before replying. Then he began in a quite unexpected way.

'This situation, Laxman, has been brought about by your own decision to keep everything here a secret. Knowledge should not be so committed to a small minority, especially if it benefits your entire species. Why don't you share it with others? Why keep them out?'

It was Arul who answered. 'Your argument, Vaman, is sound in principle, not in practice. I very much doubt that Shulz, Yamamoto and company are exactly the right recipients for Guru and yourself. They are after money and the power it brings. Guru will suit their purpose very well …'

'But why shield me?' asked Vaman. 'Guru will do as instructed. But I can think. Don't you trust my judgement of what is good for mankind and what is not? Give me the key instructions that would enable me to make copies of myself and you will have more robots like me to help you … Yet you refuse to do that.'

So they were back to the point being discussed originally. Arul replied: 'I never said we refuse, Vaman. Only, we want to go slowly about it. As you know, we humans suffer from indigestion if we over-indulge ourselves with rich food. A sudden input of rich knowledge can be equally harmful. In fact, more so. Be patient as we go about it at our own pace.'

'I must say I am unimpressed by your logic, Arul. Look where it has brought you. With such potential stores of knowledge as Guru and me, here you are confined within these walls … and Urmila, an innocent victim of it all, languishing inside', Vaman said. 'While Shulz and company roam at large.'

'Well spoken, Vaman. You have more practical sense than these two geniuses.' Urmila added her own warm support.

'Ha! So you two are conspiring against us', Laxman said jokingly. 'Be it so. But Vaman, however irrational we may be by your standards, you have to bear with us. So if you want to solve Urmila's problem you have to do so under the existing conditions. To begin with, remember Urmila cannot go out because there is a direct threat to her.'

And Laxman described to Vaman how Balu had been spotted in the neighbourhood, shadowing Urmila. He also described Balu's criminal record, and added: 'Much though I sympathize with my dear wife, I entirely share Major Samant's apprehensions. Indeed, both Arul and I agree that behind Balu is Shulz, tucked away somewhere.'

'Just as it was Shulz who motivated Pyarelal's death', Arul added by way of his support.

'Shulz … it is Shulz everywhere. Why don't you arrest him?' Vaman asked.

'Karl Shulz is wanted by many countries, but there is no evidence solid enough to ensure his confinement behind bars … That is the trouble. And, of course, this time none of the official entry points to this country have informed the Major that Shulz is in India. The Major has a hunch that he has somehow managed to smuggle himself in.'

'Arul's explanation leaves me with but one course of action to recommend', Vaman replied. 'By confining us all to these walls, the Major is playing a defensive game. It is like confining yourself to your house because it is raining outside. I would rather venture out with an umbrella.'

'What do you mean, Vaman?' asked Laxman.

'That, as you put it in English, attack is the best form of defence. I suggest we let Urmila out on her shopping expedition as before, but keep a discreet watch and catch Balu and company red-handed. Who knows … we may even draw out Shulz if we are lucky.'

'I cannot allow Urmila to be a decoy, if that is what you mean', Laxman said immediately.

'Laxman, what I am suggesting may possibly involve Urmila in a worrying experience; but I can assure you of her complete safety.' Vaman's voice was even as usual, but the words exuded complete confidence.

'How?' Laxman was unconvinced.

'There are two reasons. First, even if Urmila is kidnapped, she will be used as a bargain for Guru and me. Shulz values us much more than he values her. Apart from this purely psychological reason, I am going to make a tiny transmitter which I will attach to one of her molars. No matter where she is taken, we will be able to locate her … the device will be accurate to one centimetre.'

Laxman could not fault Vaman on either of these points. Nevertheless, he made one more attempt to persuade him.

'Vaman, as I see it you are promising me a series of safeguards. First that our own security will prevent kidnapping. Second, if Urmila is kidnapped, your device will lead us to her. Third, that you don't expect Shulz to kill her so long as he is promised an exchange … Urmila in place of the information on Guru's CPU and yourself … But suppose we find Shulz ready with a gun pointing at Urmila when we do locate her …'

'Leave it to me, Laxman. I know how to handle a terrorist. Remember, deep down in his heart Shulz is a mercenary. He is not the do-or-die type terrorist motivated by some high ideal … However, let us discuss my plan with the Major who, I am sure, will understand me better … Sister … are you willing to go through this ordeal should it become necessary?'

Urmila, who had been silent so far, now spoke up decisively: 'I entirely agree with your solution, Vaman. Any ordeal would be preferable to my present predicament.'

'Well spoken, Urmila! I am confident that you will soon be free of it … Come, my friends, let us seek out the Major.'

As they followed Vaman out, Laxman could not help noting how Vaman had changed. In an unobtrusive way he was becoming the decision-maker. It was good to have him as an ally.

6 The Kidnap

Having finished her shopping, Urmila was looking for a scooter rikshaw. This was her first outing for several days—Vaman had indeed succeeded in convincing Major Samant that his plan would work. She would normally have taken the jeep, but it was out of order. So eager was Urmila to venture out, that she insisted on using a hired three-wheeler. The Major had reluctantly agreed, but only on the condition that she would be shadowed by Jeevan in another three-wheeler. Not that the Major expected anything to happen today—for no one beyond the Science Centre could possibly know that she was venturing out after so many days. Tomorrow, of course, it could be another matter.

Certainly, today things had gone well so far.

Urmila ignored the first two free rikshaws, just in case. This had been Major Samant's advice. As she got into the third one, Jeevan hopped into another just behind.

'The subject on her way to Science Centre in rikshaw number …' Jeevan radioed the information to the communications centre where Major Samant was personally listening in.

'Follow at a distance of no more than fifty metres', the Major's crisp voice crackled into Jeevan's ears.

'Instructions received. Roger.' As he signed off, Jeevan leaned back to relax. Today was essentially a rehearsal and things were proceeding normally. Of course, with a jeep it would be better since he could ride in the back.

Soon they moved out of the congested town centre and Jeevan could see the small lane branching to the right, off the main highway, the lane that led to the Science Centre.

'Turn right here' Urmila ordered as the rikshaw approached the turning. But the driver shot past the turning.

'I say, turn back and go along that lane over there!' Urmila shouted, but to no effect. The driver continued along the highway, accelerating his vehicle to full speed.

Jeevan sprang into action and spoke into his walkie-talkie. 'Major Samant … the subject's rikshaw went past the turning to the Science Centre and is rushing down the highway. I suspect a kidnap …'

'Follow her … I will join you in the jeep.' The Major rushed to the door where the jeep, now repaired, was standing ready.

'Damn!' he shouted a minute later. The rear left wheel showed a flat tyre. And the driver was nowhere in sight.

Major Samant was not one to heistate long. His own motor cycle was parked nearby. He hopped onto it and raced away, making sure, however, that his favourite revolver was in his pocket.

'*Pinnale Poi … Walatupakkam Tirumpu.*'

'*Hinde Tiragi … Balagade Hogu.*'

Urmila's instructions in Tamil and Kannada fell on deaf ears and she finally realized that things had gone wrong. She leaned sideways and looked back. Another rikshaw was following. Probably Jeevan was in it. She abandoned the thought of jumping out of the vehicle as she saw the hard concrete on one side and a ditch on the other. In any case, the Major had told her in no uncertain terms not to attempt anything foolish.

Meanwhile, Jeevan found that his rikshaw was falling further and further behind. As he enjoined the driver to go faster, he replied, 'Sir, that is impossible. I am going at full speed and can go no faster than this. The rikshaw in front has some kind of booster to increase its speed. Normally these vehicles don't go that fast.'

Jeevan saw Urmila's rikshaw overtake a truck. As he approached the truck he found that it was, in fact, stationary and had lost one wheel. When his rikshaw was about to shoot past it, another truck came from the other side and stopped beside the stationary truck. The two drivers appeared to have embarked on a long conversation in Kannada about the breakdown. Meanwhile, the road was totally blocked.

'Move to the side, you fool!' Jeevan shouted as his driver blew his horn continuously.

'All in good time … let me find out what is wrong here first,' the truck driver replied insolently.

'All in good time—you will be in police custody if you don't back up instantly,' Jeevan displayed his ID card as he said this.

That sobered down the driver, who reversed his truck … but not as promptly as Jeevan would have liked. He noted the registration numbers of both trucks as his rikshaw once again sped on. But four precious minutes had been lost and Urmila was by now out of sight. Still, Jeevan continued along the highway until they reached an intersection. Which way had Urmila gone?

'Turn left', Jeevan ordered. He could make out some faint marks on the dirt border where, he guessed, Urmila's rikshaw had made a sweep to the left.

Jeevan's guess turned out to be correct, for he saw a rikshaw coming towards him. But the hopes he entertained were dashed as he saw the empty passenger seat. And the driver's head was bleeding.

As Jeevan stopped the rikshaw, he was conscious of a motor cycle screeching to a halt behind him. Major Samant had arrived.

For good measure, Jeevan slapped the driver and asked, 'Come on, tell us where you took the lady.'

'What lady, sir? I am unable to think properly. I drank too much and kept driving until I bashed my head against a tree, sir!'

Don't lie, you scoundrel! I can tell a drunk when I see one.' To reassure himself, Jeevan smelled the driver's breath and then slapped him again. 'Come out with the truth, or you'll get such a bashing at the police station …'

Major Samant interrupted the squabble in a calm and authoritative tone. Realizing that Jeevan was upset at losing sight of Urmila, he said 'Kidnapping well get you a long jail sentence, driver. The least you could do to reduce it is to come clean now. Tell us quickly, where did you take the lady?'

The driver broke down and tearfully confessed: 'Sir, this is the truth, I swear on my life. I took the lady about a kilometre from here to a waiting car. A white Ambassador, sir … Oh god … punish me for my greediness that I did this for a hundred rupees … I was told that it was a rehearsal for a film. Believe me, sir …'

But Major Samant was already on his way, having instructed Jeevan by a sign to take the driver into custody.

About three kilometres further on he saw a white Ambassador parked by the roadside. It was empty. There was a footpath leading from the spot to an empty field. As he looked towards it, Major Samant bitterly cursed to himself. A helicopter was rising into the sky, going due east.

He was brought back to reality by the arrival of his own jeep. The Major kept his voice under control as he asked the driver:

'And what made you abandon the jeep?'

'Why, your message, sir! You asked me to bring Navin Sahib to you', the driver said, surprised.

'I asked you? When and how?' The Major was equally taken aback. The driver fumbled in his pocket and produced a piece of paper torn from his memo pad. The Major read the message on it:

Driver Sharma,
Please bring Navin Sahib to me immediately. He is in his barracks.
Major Samant

The handwriting and the signature below were unmistakably his. And there was also his official stamp underneath.

7 The Ransom

The phone rang at eight in the night.

Although a call had been expected, the shrillness of the bell brought everybody to attention. It was Major Samant who lifted the receiver.

'Major Samant speaking', he said. He knew that the call was from the kidnapper. A handwritten note delivered late in the afternoon to the security guard at the Science Centre had warned them to expect a phone call at eight.

'I want to talk to Dr Laxmanan', the voice had a foreign accent. Samant motioned to Laxman, who picked up another receiver. The recordist tapping the conversation was already busy with his work.

'Laxman speaking.'

'This is to assure you, Dr Laxmanan, that your wife is safe in our custody …' The voice at the other end betrayed no emotion.

'Who are you? Where are you speaking from, you wretch?' Laxman was beside himself.

'Pacify yourself, Dr Laxmanan. You have to listen carefully to what I say next … your wife's safety depends on it', the voice continued.

Major Samant, who was listening in, patted Laxman on the back. Laxman nodded to indicate that he had regained his self-control. Both listened intently.

'In exchange for the safe return of Urmila, I need the following information by two o'clock tonight.' The voice continued, 'I need the correct specifications of Guru's CPU—I repeat, correct details … please don't palm off false information any more. And I need Vaman delivered to me … How the exchange is to take place will be communicated to you at one o'clock … that is, at one hour past midnight. Be sure to take all this seriously … or else.'

'Or else, what? Speak up, please!' Laxman shouted. But the click at the other end told him that further shouting would be futile.

The recordist, who was listening carefully, played the tape once again before giving his opinion: 'It is Karl Shulz all right.'

Samant issued instructions to locate the source of the call. But Laxman now broke down completely. He shook Samant by the shoulders and said through sobs, 'So much for your security! The tiger has snapped the decoy … that is all you think about … but who is going to bring back Urmila? Tell me that … Will anybody here tell me that?'

There was silence in the communications room. Even Samant had nothing to say. But from near the door, a mechanical voice was soon heard distinctly.

'I will tell you, Laxman. The responsibility is mine. Come, let us locate Urmila.'

As everyone looked towards the door they found the tiny figure of Vaman standing there.

The helicopter carrying Urmila descended on a field beside the national high-way linking Bangalore and Hyderabad. Although it was dark, Urmila could make out something that looked like a farmhouse standing some distance away.

'Be my guest tonight, Mrs Laxmanan.' Shulz bowed as she stepped onto the ground. She had already recognized the other person in the helicopter as the notorious Balu.

'Please follow me.' Shulz led the way with a tiny torch, with Balu bringing up the rear. There was a narrow footpath amidst overgrown grass, and soon Urmila found herself at the farmhouse.

It was a dilapidated structure with no windows and two doors, both of which were closed. Balu opened the one away from the highway to reveal a room lit by a lantern. Urmila saw that something like a meal was laid out on a table, a rickety chair beside it.

Two rough characters suddenly materialized out of the dark. Balu's men, she surmised. She was right, for Shulz spoke in his quiet but menacing tones: 'Balu, you and these pals of yours had better guard the lady well. But no rough work with her, mind you … or I will skin you alive.' Even those hard-ened criminals paled at that warning, Urmila noted. Then Shulz turned to her and said politely. 'Madam, it now depends entirely on your husband. If he is prompt with what little I have asked for, your stay here need not be even until dawn … If he is foolish enough to try to cheat me, I'm afraid you will not see dawn.' He pointed towards a corner of the hut.

Urmila looked and shivered. A rectangular pit two metres by one metre, and about half a metre deep had been dug there. When she turned round Shulz had disappeared.

Everyone followed Vaman to a computer terminal, to which a small gadget had been attached.

'I made this little toy', Vaman explained. 'It will collect the signals and, with Guru's help, process them to tell us where Urmila is.' A map of India had appeared on the terminal.

'Signals? From where?' asked Navin. He was one of the crowd.

'From the little transmitter I had fixed in Urmila's molar ... that was my secret weapon which only three people here know about', Vaman said.

'Well, I am not amongst the favoured three' Navin retorted. Major Samant thought Navin looked a little uneasy. Why, Samant wondered? Or was it his imagination?

But he was distracted from his thoughts by what was happening on the terminal monitor. A green light, like a tiny dot, had appeared within the map of India. The dot was flashing with intensity.

'That is where sister Urmila is', Vaman pointed to the dot triumphantly. 'I could not get it earlier because she was airborne, in a helicopter, and because my transmitter does not act very efficiently in mid-air. But now that she is on the ground, I can locate her. She is stationary now.'

'But where exactly is she?' Laxman asked with desperation.

'We will soon find out by changing the scale of our map', Vaman replied. He enlarged the scale so that first only the southern parts of India, then the border between Karnataka and Andhra Pradesh, and then the national highway from Bangalore to Hyderabad appeared on the screen. The green dot continued to flash steadily.

'Your maps of the region are not worked out in great detail, otherwise I could have told the location within the error of one centimetre. As it is, I can only tell you the approximate location with regard to the nearby villages and the national highway.' Vaman pointed to two villages on the highway, and continued: 'According to my calculation, sister Urmila is about a quarter of a kilometre away from the national highway and at a spot which lies ten point three five kilometres from the southern village.'

'Good enough for me! I will now arrange a helicopter and a crack group of commandos who have been on standby ever since we decided to go ahead with this project', Major Samant said with enthusiasm.

'Then why don't we get Umi out before one o'clock? Let Shulz shout himself hoarse after that.' Laxman was also infected by the new optimism.

'Take it easy, all of you!' Vaman cautioned. 'Your helicopter will be noisy and alert whoever is keeping guard on Urmila.'

The Major wiped his perspiring forehead. This tiny robot was now teaching him strategy. But the little figure was right. Aloud, the Major said, 'We will drop the task force ten kilometres from the spot and let the rescuers cycle to the spot.'

'That is well thought out, Major', Vaman applauded. Then he added, 'And I suggest that you mount the rescue operation after one o'clock.'

'Why? Wouldn't sooner be better?' Laxman asked impatiently.

'Vaman is right, Dr Laxmanan. It is now clear that Shulz is not where Urmila is. Figure it out youself. He will call us at one and we will have to deliver the goods to him within the hour. So he won't be far from here. If we rescue your wife after one o'clock, he won't know that his operation has misfired. We will then be able to catch him too, red-handed. If, on the other hand, we rescue Urmilaji much sooner, he will get suspicious and disappear altogether … Dr Laxmanan, you have already done so much for us; may we request you to allow us this extra time?' Major Samant asked as softly as he could manage.

Somewhat reluctantly, Laxman nodded agreement. Arul whispered in his ear, 'A brave decision, Laxman. All of us are beholden to you.'

8 The Mole

The electronic clock in the communications room showed seconds ticking away. To Laxman the time taken to go from 00.59.00 to 01.00.00 seemed like ages. He expected Shulz to be punctual … so methodical had he been until now.

The phone rang at 01.00.04. At Samant's nod, Laxman picked up the receiver and identified himself. Because the Major had attached a loudspeaker to the phone box, everyone in the room could now clearly hear each word that Shulz spoke.

'Dr Laxmanan, listen carefully. In the drawer of the cash register of your canteen you will find a yellow sheet of paper. It has a map of your locality with two points, A and B, clearly marked. At a quarter to two, Vaman must appear at point A along with the information I asked for. The information has to be in the form of the package you got from the container—not an interpreted or doctored version. Let me warn you, I have means of checking whether the version is genuine. It had better be, for your wife's sake! The car which brings Vaman to A must return right away, after flashing the headlights twice. And after fifteen minutes Vaman should walk to the spot marked B to be picked up … is everything clear?'

'Yes … but what about my wife? Where is she?' The Major smiled in approval at the desperate note of anxiety in Laxman's voice. Karl Shulz must not know that Urmila had been located and would be rescued shortly.

'All in good time, Dr Laxmanan. If you fulfill your side of the bargain, trust me to do the same. At six in the morning, you will know her whereabouts.'

The loud click informed everyone that Shulz had signed off.

By now a security guard had retrieved the yellow sheet, found exactly where Shulz had said it would be. Major Samant took it as another addition to the mounting evidence that Shulz had an accomplice right there in the Science Centre itself. Otherwise, how could Urmila be kidnapped on her very first outing? Who had waylaid the jeep driver with a false note and flattened one tyre?

'Let the experts look for fingerprints right away!' Samant stole a glance at Navin as he said this. Navin himself was staring at the yellow sheet.

'I brought it along carefully, sir.' The guard had used gloves while holding the paper. But he had more to add. 'It is strange, though, that the paper has a perfume.'

'No doubt used in lovers' correspondence … we will soon find out', Samant added drily as he sniffed the paper before passing it on for examination.

It was now five past one. He had to give marching orders to the commandos. He walked over to the radio phone and spoke: 'Operation Umi: Come in.'

'Ready for action, sir', the reply came in loud and clear.

'Go ahead.'

The commando task force had, in fact, reached the intended spot by midnight. First by helicopter and then on bicycles, the commandos had completed their journey unobtrusively. Their last lap on the highway was completed without lights. Fortunately for them, the traffic on the road was thin. If any truck driver had noticed a bunch of cyclists riding without lights, he had not reacted to that sight.

'Stop', the leader of the unit ordered, himself slowing down. He had spotted the farmhouse. But to be doubly sure, he examined under torch light the location printed out by the computer. 'This is it, men … let's go and wait over there.' He pointed to a spot about fifty metres from the hut, where the tall grass provided good cover. The five commandos were soon swallowed up in the darkness.

'Vaman … don't you need some of us to come with you?' Laxman asked anxiously.

'No, Laxman! We don't want to make Shulz suspicious—at any rate not while Urmila is still in his custody. And don't worry about me … I can take care of myself with this little toy', Vaman replied.

The 'little toy' was a laser gun that Vaman had made during the day. He demonstrated it on a metal target and even Major Samant was impressed.

'Yes, I can take care of myself even though our friend is tall and hefty. Have you forgotten the story of David and Goliath?'

Laxman was not surprised that Vaman knew about this tale from the Old Testament. Nothing would surprise him about Vaman's ability any more. As Vaman got into the car, Laxman pointed to the briefcase in his hand.

'Make sure that Shulz does not suspect that he is getting false information again.'

'Have no fear! Shulz will have no opportunity to suspect', Vaman said.

Major Samant had one parting order for Vaman. 'Don't be overzealous with your toy, Vaman. We want Shulz alive.'

There was a pause before Vaman replied. 'I don't expect the need for the gun will arise.'

As he left for the rendezvous, Major Samant looked at his watch. It was a quarter past one.

'Please, Major Samant, can you turn the light away?' Navin was in great discomfort, but his appeal had no effect on the Major, whose face was set hard as granite.

'I have to get the truth out of you, Dr Navin, and sometimes I can read it better in a face than when I hear it spoken. The light helps.'

Earlier, when Major Samant had politely sought a word with Navin, the latter did not expect to endure what seemed to him 'third degree'. Samant had taken him to a closed room and seated him at a table with his face barely a foot away from a table lamp with a 100 W bulb in full power.

'Let me congratulate you, Dr Navin, for your deception.' The Major spoke from a corner where he could not be seen. 'I thought that you had turned over a new leaf and were cooperating … Tell me, when exactly did you start helping Shulz again?'

'I swear to you, Major, I have not helped Shulz on any occasion since the day I promised to go straight', Navin said with passion.

'Ha! Swearing will get you nowhere. This paper with your fingerprints speaks for itself.' The Major showed him the yellow, scented paper and added. 'The map is drawn with a green ball pen like the one you possess and which nobody else has here. The letters are in your handwriting, so my experts assure me.'

'I admit the paper is mine. The pen could also be mine … but I'm baffled by your evidence. I am sure it is fabricated. Someone is trying to frame me', Navin remonstrated shrilly.

'Seventy per cent of the criminals caught red-handed say so. Let me give you my version. You were dead scared when Shulz killed Pyarelal. Scared enough to cooperate with him again. And with your plans for matrimony with the socialite Miss Runa, I am sure you needed money too … it all fits.'

A series of loud knocks at the door interrupted Major Samant. He opened the door to be confronted by Arul and Laxman, both highly agitated.

'All is lost, Major', Arul blurted out. Major Samant suddenly felt deflated.

'Tell me quickly, is Urmilaji …' he began apprehensively.

'Not Urmila. We are still awaiting news of her.' Laxman added, 'It is about Vaman. A great mix-up has occurred. He was to take a package of false data to Shulz. It was all packed in a briefcase. I saw him take it with him …'

'Yes, Dr Laxmanan, I too saw him take it.' The Major was now somewhat calmer in mind.

'But that was the wrong briefcase, Major! Someone put the correct package in Vaman's briefcase. And the false one is still here in another identical briefcase', Arul said as he produced a briefcase.

'Yes, the original, correct package is gone', Laxman confirmed. 'So Shulz has got what he wanted, thanks to this mix-up.'

'Where did you find this case?' the Major asked.

'Lying on Navin's table. In fact we discovered it accidentally, as we went to look for him there. We cannot understand how it happened Major.'

'I do. It all fits', the Major said. Then turning to Navin, he added, 'Perhaps all this is fabricated too? Dr Navin, I have no alternative but to place you under arrest. Let me put you somewhere for the night where you cannot cause any more mischief.'

Arul and Laxman stared at Navin. The Major felt he owed them an explanation. 'What you have found fits naturally into the whole picture. Dr Navin has truly outwitted us. I will give you the details in the morning when we shall all sit in judgement on this master accomplice. He has throughout been aiding and abetting Shulz. As far as Shulz is concerned, our only hope now rests in Vaman ... Can he handle that master criminal? Come on Dr Navin, follow me.'

As Navin left the room he stared long and hard at the briefcase. To Arul and Laxman, that look communicated a deep desolation and despair.

At five past one the commandos started moving towards the hut, crawling on the ground to avoid detection. There were no bushes near the farmhouse to conceal them from watchers within, but the almost total absence of moonlight helped.

Near the house the leader of the unit and one commando went to the left and the other three to the right. A man armed with a gun and a walkie-talkie stood alert at the door facing the highway. The commandos had no difficulty recognizing him as Balu. At the other door sat another watchman, apparently half asleep. The task force had no knowledge as to whether others were inside. If there were, they must not know that their colleagues were being overpowered.

The dozing watchman had a momentary phase of total wakefulness before he passed into oblivion. However, a slight noise alerted Balu.

'Rajan, are you OK?' he asked from his end. Receiving no reply, Balu let out a curse and muttered, 'The wretch has dozed off. Let us wake him up with a whack.' He walked across and saw Rajan apparently lying asleep on the ground.

'Wake up you idiot!' he accompanied the invocation with a kick. Simultaneously, he felt a karate chop descend on his shoulder and he knew no more.

'Two down', announced the Captain softly. Meanwhile, other commandos had entered the hut and overpowered the third man, who had been fast asleep. Urmila was awake, terrified but unharmed.

'Operation Umi over and successful.' The Captain radioed the message.

Major Samant gave a sigh of relief and looked at the clock. It was one thirty.

Exactly at a quarter to two a car dropped Vaman at point A and turned back. But it did not go straight back to the Science Centre.

About a kilometre from point A, it turned into a side lane and stopped. The driver opened the boot and let out two commandos. Then the car turned back and reported to the Science Centre.

The two commandos made their way swiftly but unobtrusively to spot A. By then Vaman had begun his solitary walk to spot B. The commandos hid themselves behind a bush and watched his progress.

9 The Document

The place selected by Major Samant as a temporary lockup for Navin was the same large hall in which the container and its contents were well guarded, the door to the hall was fitted with a formidable lock. As the key turned in it, Navin became agonizingly aware of his lonely confinement.

He had tried his utmost to convince Samant of his innocence. But when he asserted that the evidence was fabricated and planted, the dour Major had turned round and asked, 'Whom do you suspect?' Indeed, to this question Navin had no answer. He had lost contact with Shulz long ago, and now Pyarelal was dead. The way all the incriminating evidence had turned up pointed to an inside job. But who, in this Science Centre, wanted him out of the way so desperately? Since nothing would happen now till the morning Navin pushed this nagging question away and looked for a place to sleep. It was then that he noticed the plaque on the solitary bench in that large hall.

It looked familiar ... where had he seen it before?

Then memory came rushing back. He recalled Arul telling him that the plaque was found while digging, but that it was at a shallower depth than the container itself. It had been kept aside and eventually forgotten with the discovery of the container's sensational contents. Navin remembered seeing the plaque and the letters on it. At that time he had not mastered the key to that ancient script, but he could now read it as easily as a modern alphabet. Looking at the letters again, the archaeologist in him took precedence. Perhaps this plaque was intended to exhort the digger to go deeper and find the container. With mounting interest he turned his eyes over those strange symbols, his mind piecing together their message.

The plaque contained words in bold red letters in the central square, surrounded by a considerably longer text in a smaller, black script. Reading the red-lettered message alone, he sat up, startled. He then read the text in black as fast as he could.

When Navin had finished, he realized that a message had to be communicated urgently. It was essential to get in touch with Major Samant. He looked round desperately … and the sight of a phone on the barricaded window lifted some of his gloom.

'I must talk to Major Samant, please.'

'Samant, here … Of course, it's Mr Navin, is it not? Getting tired of your confinement already? Or do you have a confession in mind?'

'Major, please! This is no joking matter. We have been taken for a ride. I have now solved the mystery … and must explain it to you all … please bring Arul and Laxman too.'

'All right … but let me caution you: no tricks.'

'Bring an army to watch over me if you like … but I beg you, don't lose a moment!' Navin's voice was desperate but sincere, and Major Samant acted quickly.

Navin was clearly relieved to see Arul and Laxman with Samant. As soon as they entered he led them to the plaque.

'Remember this plaque?' he asked. Arul and Laxman nodded. They too had ignored its existence once they were absorbed in the mysteries of the container. Navin continued:

'This is where all the relevant information is noted down … we were fools to ignore it.'

Major Samant intervened. 'Mr Navin, I though you had hit upon some new evidence. This plaque is thousands of years old. What possible bearing could the inscription have on our problems today? We want to know about Urmila's kidnapping, the ransom we had to pay in the form of Vaman, and about the mole who is giving our secrets to Shulz and company … I have no time to waste on ancient inscriptions. I am anxiously waiting for Vaman to return.'

'You may wait till the cows come home, Major Samant!' Navin smiled ironically. 'Vaman will not come back. And, by letting him go and shutting me up here you have committed the classic confusion between the guilty and the innocent.'

'Take it easy, Navin. Surely you don't imply that Vaman is guilty?' Laxman asked incredulously.

'The evidence for Vaman's guilt is written here, gentlemen. Let me read it all out to you straightaway.'

And without more ado, he began with the inscription in red:

'Beware! Those who happen to discover the container beneath, be sure to read first what is inscribed herein.'

Navin then continued with the inscription in black:

'Should you discover the container and be clever enough to decipher its store of knowledge, please exercise the utmost caution so that you avoid the fate that befell us …'

The three listerners quietly heard the denouement read out by Navin in sepulchral tones.

When the container was buried, we Monads were riding high on the crest of prosperity. We were able to draw upon the ample reserves of this beautiful planet to sustain our great civilization. And we were confident of being able to do so for a long time to come. Nevertheless we decided to enclose information about ourselves in a time capsule so that, should unforeseen events extinguish our civilization, those records would tell those who follow us, possibly centuries or millennia later, how we had flourished. Little did we know that the end would come so soon.

But such is the situation now. Within ten days of the internment of this plaque not a single Monad will be left on this planet. The tragedy is that the end did not come from any natural calamity but was brought upon us by our own actions. Lest you who discover this container are tempted to do the same … be warned.

The container will give you information about our science and technology, our music and fine arts and about our philosophy. You may be dazzled by it and may want to follow in our footsteps. But, friends, be careful how you use this information. Let our fate serve as a warning.

We recall the day when our scientists achieved the zenith of progress, when they were able to create and employ artificial intelligence. A thinking but mechanical Monad, one who could make a copy of itself. This achievement not only had tremendous practical uses, but also boosted our ego. We had finally succeeded in creating something that, like us, could think and reproduce … we called our mechanical counterparts *Konads*.

How we celebrated that day! The teams of scientists who had perfected the Konads were feted all over the world. And we planned the numerous ways in which Konads would serve us, to make our life even more agreeable.

The Konads translated all our expectations into reality. They solved with great ease the outstanding problems that had bothered our scientists. They implemented the plans on the drawing board of our technologists. With their intelligence they rapidly improved themselves, becoming increasingly more knowledgeable and efficient. They were evolving.

That was when we thought of a time capsule … of the container beneath that provides the key to all our achievements, including the making of

Konads. The container, as we now realize, was motivated at least partially by the Monadic ego.

Little did we realize that we were also evolving, but down the spiral of progress. As our dependence on the Konads increased we began to lose our zest and initiative. Why work if everything we wanted done was accomplished so much more rapidly and efficiently by Konads? As their masters, we could command them to do our bidding. So why work?

The Konads were watching our slide into inaction. Our brains, which had evolved to their present state by facing so many challenges, now went into a state of stupor, while the Konadic brain was improving rapidly. And their numbers were also increasing—something we did not realize until it was too late.

The Konads made us believe that by multiplying their population they would serve us so much better. But in reality their sole motive was their own survival and welfare. They were willing servants and pupils so long as we could teach them something. When that stage was past, we became redundant in their scheme of things. And, of course, their super-efficient brains told them to discard everything redundant.

So they decided to do away with us.

That terrible morning is fresh in our memory. All the Konads went on strike. They simply disappeared from our colonies. And that was enough to cause chaos. Food, energy, medicine, entertainment—everything we needed had come to depend so much on the Konads that we were rendered totally helpless. We called upon the Konads to come back and resume work, we offered them inducements, we threatened them, we even tried to divide them. But nothing worked. And we could see their game. Merely their inaction would bring about our end. They did not need to waste energy on violence.

And they were right. Already, as we write this, the Monad population on this planet is down to half its original size. Lacking our forefathers' fitness to struggle for survival, the rest of us will succumb within ten days. But we have one cause for satisfaction. We have arranged for an automatic and irreversible shutdown of all our power houses. So when we go, the Konads cannot survive much longer either. After all, they too need energy.

How we wish we could destroy all the information in the time capsule that deals with the making of Konads. But we neither have the means nor the energy to dig and recover the container. So we leave this plaque as a warning.

In the darkness of despair around us, we see one ray of hope, however. Although all our communities are going to die here, that is not the end of the human race. Some primitive tribes are left on this planet. Like other animals, we have left them untouched in the jungles. They were part of our experiments on evolution under natural processes. So we never interfered with their existence. Our hope is that one day those tribes will progress to a stage when they will be able to appreciate all we have done.

To those descendants of our primitive brothers we leave this warning. Do not bring Konads into existence. At first they will seem friendly. But they are not friends. They are not enemies either. Their sole aim is to better themselves, without regard for anyone who stands in the way. We entreat you, therefore, do not allow Konads to come back to this planet.

So, beware …

The Monad Society

10 Farewell

There was silence as Navin finished reading, with those present attempting to adjust to the startling revelations from an extinguished civilization.

Finally, Arul spoke: 'It was a great mistake to have overlooked this plaque … we were so hopelessly carried away by the container and what it revealed. Otherwise, we would not have developed Vaman …'

'I just can't bring myself to believe that Vaman is a villain', broke in Laxman. 'Look how he put all his efficiency at our disposal in order to locate Urmila.'

'Precisely the pattern the Konads followed in their previous incarnation—if I may use the word!' Navin's face wore a bitter smile as he continued. 'Like the Konads, Vaman made sure that he had our confidence—that in our eyes he was a hero.'

'I tend to agree with Mr Navin', Major Samant commented thoughtfully. 'It was Vaman who suggested that Urmila be allowed to move freely. I now suspect—no—I am sure, that he was regularly in touch with Shulz. He conveyed a message. A message to Shulz to say that the coast was clear for his operation.'

'Yes, it all makes sense now. Because everything was pre-planned, Urmila was kidnapped on the very first day that she ventured out … and Major Samant's driver was sent away with a false message and the tyres of his jeep were flattened … and later, evidence was fabricated against Navin—Vaman could easily have done all this', Arul added somewhat breathlessly, now that the real culprit was being identified.

Laxman was still in a state of shock. 'But why should Vaman do such things? He was amongst friends here—he had the run of the place …'

'A Konad has no friend, Laxman', Navin broke in. 'True, you and Arul were friendly with Vaman. You gave him the ability to think. But you were hesitating to give him the power to reproduce. That power would be readily available to him, courtesy of Yamamoto & Co. So he has now chosen to go to them. All his steps are carefully worked out as a part of a plan to multiply and conquer Earth.'

'Which is why Vaman made sure of taking the real stuff in his briefcase.' Major Samant looked at his wristwatch. 'Quarter to three ... our David appears to have absconded with Goliath.'

There was a sudden beep on the portable receiver tied to Samant's belt. Motioning for silence, he turned up the volume of the receiver.

'Commando Shersingh here, sir. We were ready to help Vaman in his fight, as arranged. But he offered no resistance to Shulz. In fact, he went eagerly with him. We tried to stop them ... but both Ramsingh and I have been immobilized by Vaman's laser gun.'

'Where are the two of you?' Major Samant asked in his crisp tones.

'At place B on the map, sir! But please follow Vaman and Shulz ... they are taking off in a helicopter right now ... over.'

'Let's be on our way, gentlemen.' Major Samant dashed out of the hall. 'I hope the helicopter bringing Urmila is back now ... we may still be able to persuade Vaman to come back.'

Samant was right. The helicopter was about to touch down as all of them emerged from the building. Laxman rushed ahead.

'Umi darling ...' He almost smothered Urmila as he gathered her in his arms.

'I'm fine, Laxman ... please thank these commandos for rescuing me alive. But how were you able to locate me?' Urmila's voice was very unsteady, but also very happy.

'With Vaman's help! Remember the tiny transmitter he fixed in one of your molars?'

'Then let me thank him first.'

'That, I'm afraid, is not possible.' Laxman laughed mirthlessly. 'The little fellow has changed sides. He ran away with Shulz.'

'Vaman ... in league with that horrible man? I can't believe it!' Urmila exclaimed.

'I wouldn't have myself, but for cast iron proof ... but I must leave you darling and go after them.' Laxman gave her a hug and dashed up into the helicopter which was now ready to take off with Major Samant, Arul and a few commandos inside.

'Take care!' Urmila waved. How much she wanted Laxman with her now, but obviously his errand was important.

'Don't worry Urmilaji ... These commandos and I will take care of Shulz while your husband and Mr Arul give the little fellow a real talking to.'

As the helicopter rose Laxman said, 'If the little fellow is in no mood to listen, I will have to use my ultimate weapon.'

'Which is what?' asked Major Samant.

'I can inactivate his energy source by remote control. That will immobilize him for sure.'

'Don't be so sure, Laxman!' Arul said, bringing out a note and a packet from his pocket. 'Here. Have a look at what I found on my table just before I came here.'

The note was addressed to Arul. 'Dear Arul, please pass on my parting gift to Laxman. I don't require this toy which he made specially for me … Yours, Vaman.'

As Laxman read the note he felt the 'toy'. Even before he saw quite what it was he had guessed correctly. It was the same microwave receiver that he had fixed in Vaman's brain. So Vaman knew not only of its existence, but also of its purpose. Which was why he had sent it back.

Why did he write the note to Arul? Why not to Laxman himself? Did he have a sense of guilt towards Laxman? … Then Laxman realized that Vaman possessed no conscience. All he did was carefully reasoned and calculated to improve his own future prospects. Vaman had addressed the note to Arul purely because his office happened to be near the exit, unlike Laxman's at the back.

Arul glanced at Laxman. His face was frozen, with no trace of emotion. Was he feeling betrayed by his protégé? Arul then felt in his pocket for the ultimate weapon—that was made under advice from Professor Kirtikar and still rested there. Would Vaman have guessed its existence too?

Meanwhile, the helicopter was speeding along a flight path selected by Major Samant. About a hundred and fifty kilometres south there was a disused airstrip. Samant suspected that Shulz would have a small plane ready and waiting there.

'How far can he go in a small plane?' asked Arul.

'Far enough!' replied Major Samant. 'The bastard will most likely land in Jaffna where things are in turmoil. From there he may have made further arrangements … The important thing is to catch him on our soil.'

In half an hour Samant's suspicions proved correct. A bright dot appeared on their radar screen. Their helicopter was new and had a powerful engine, while the machine acquired by Shulz belonged to an older generation.

'That is our hope for overtaking him … but I would rather catch him on the ground.' Samant instructed the pilot to shadow the fugitive helicopter without approaching too close. He then began to brief the commandos. That was when Arul turned to Laxman.

'Laxman, I want your permission for a certain course of action, should it become inevitable.'

'What the hell are you talking about?' Laxman asked peevishly. He was still not reconciled to Vaman's treachery.

'Pull yourself together, Laxman,' Arul uncharacteristically spoke with anger. 'I know how you feel about Vaman. If we are not able to persuade him to come back to us on our terms ... do I have your permission to use the final step?' Arul now asked, gently but firmly.

'What final step? Shulz and Vaman hold all the aces ...'

'When the opposition holds all the aces it is time to quit the game ... if Vaman cannot join us, we must ensure that Shulz cannot have him alive ... I have made such an arrangement.'

'You? How and when? You just saw how neatly Vaman dealt with my own so-called preventive action. How sure are you that he has not anticipated you too?'

Arul now laughed. 'Laxman, you tried to beat Vaman at his own game. You fixed a mechanical device in a mechanical brain and you lost. I have proceeded somewhat differently. I took advantage of Vaman's ego. Yes, his ego! His realization that he possesses a superior brain. Do you recall how I presented him with my ring when he solved that problem?'

'Yes indeed! I remember how proud he was at his achievement. That he should have solved a long-standing problem within two days of acquiring intelligence was a great achievement for him. How proudly he displayed that ring ...' Laxman paused and added, 'You don't mean that ring ...'

Arul nodded and asked, 'So do I have your permission?'

Laxman hesitated ... and then replied. 'Yes, but only when I give up will I signal to you.'

Arul pressed Laxman's hand in sympathy.

Their helicopter started descending. They could make out something on the airstrip. Yes, Shulz's helicopter had landed behind a small plane.

'Welcome, gentlemen!' As they landed, they faced Shulz and Vaman. Shulz's rifle pointed at them while Vaman held his laser gun.

'Karl Shulz! Under Indian law I am placing you under arrest. Throw your gun down and walk towards our helicopter, hands up in the air', Major Samant called out. Shulz laughed defiantly.

'Brave show, Major! Twelve murders and twenty-five cases of smuggling, to say nothing of a few kidnappings ... I have to answer calls from your counterparts in so many other countries! Won't they be disappointed if I submit to you? I have decided to remain at large instead!'

Laxman addressed Vaman directly. 'Vaman, what kind of future do you see with this self-confessed criminal? Come back to us. We are friends ... we forgive you all your pranks. Remember I made you. Arul here taught you so many things … Come back, dear Vaman! This is where your future belongs.'

Vaman replied, 'Laxman, even if I were to grant all you say, I am not guided by emotions like you humans. I must act with my self-interest in view. With all your professed friendship, Arul and you are reluctant to teach me how to reproduce. Karl has promised to place all of Yamamoto's organization at my disposal to do precisely that. So I know where my future lies.'

'Come Vaman. There is no point in arguing.' Shulz sent Vaman on to the waiting aircraft and turned to Major Samant.

'Major, my rifle has a long range. So don't move till our plane is on its way.'

Shulz stepped backwards, covering them all the while with his gun. Meanwhile, Vaman had reached the plane. Samant, Arul, Laxman and the commandos were watching, as if frozen.

Shulz went in and started the engine while Vaman was still at the door, his laser gun covering the helicopter. 'Vaman! I appeal to you, come back', Laxman shouted over the din. Vaman shook his head.

'No Laxman. I will return, but not just now. Later perhaps, when I can visit you with my brethren. Meanwhile, I have Arul's ring to remind me of my friends here. Au revoir!'

'Good bye, Vaman!' Laxman shouted and nodded sadly at Arul.

Vaman had turned and was entering the cabin when Arul pressed the button in his gadget. There was a blinding flash and a roar like thunder as the aircraft broke into smithereens.

'Goodbye, Vaman, goodbye', Laxman intoned to himself as he wept.

Part III

The Science Behind the Fiction

Sci-Fi and I: Science Fiction from a Personal Perspective

Introduction

My childhood was spent literally amongst books. My father, a mathematics professor could not resist the temptation of acquiring books, whether from bookshops, by mail order, or from a door-to-door salesman. So these purchases would turn up eventually in tall shelves with glass doors and the wall space would in due course be taken over by new shelves. The shelves were too tall for me and I carry a bitter memory of a misadventure at the age of four. While practising long jumps in front of a book shelf, I exceeded my normal range with the result that my right foot crashed through a glass door. I still carry recollections of being taken to the university hospital, where a kind-looking doctor transformed into a source of terror as he produced a sharp tong-like implement and proceeded to remove glass pieces from my bleeding foot.

My acquaintance with books and bookshelves was renewed a couple of years later when I began to appreciate the treasures that books contain. It began with my mother's bed-time reading out of a story book, as my (younger) brother and I listened with growing sleepiness. At some stage she would call it a day (or night!) and stop. She would, of course resume the following night. However, as this went along, there came a point when I could not hold my suspense for the next instalment to follow. I would then start reading that book myself. And having observed my father reading some book or other, I realized that I had got (- 'inherited' is too long a word for the vocabulary of a seven year old -) my habit of reading from him.

That limitless menu of books at our house contained a variety in English, Hindi and Marathi whose range and sophistication increased with my increasing age. Starting with Winnie-the-Pooh and tales of Akbar and Birbal, by the time I reached the important stage of a teenager I had become acquainted with the likes of Jules Verne, H.G. Wells and Conan Doyle. As a genre of stories which had a scientific appearance but which referred to somewhat unusual and hitherto unseen aspects of science, I began to develop a special interest for them. Later I came to know that these were *science fiction* stories. Little did I imagine then that I myself would be writing some in the years to come.

This account being specifically about science fiction, let us first see where it stands in relation to the wider variety of science writing.

Categories of Science Writing

Literature is the expression of society, said Charles Nodier. Nodier was an influential author in the French post-revolutionary era, when he introduced younger generation romanticists to gothic literature, tales of vampires and the role of dreams in literary creation. His career as a librarian was also important in the literary role he played. His influence has been acknowledged by famous authors like Victor Hugo and Alexandre Dumas.

Across the channel, Robert, Lord Lytton, statesman and poet, asserted that a nation's literature is always the biography of its humanity. Quotes like these tell us that the literature of the time emerging from a nation gives us a flavor of its society at that time. Applying this maxim to the present times, we should expect an important contributor to a nation's literature to be its science. For science exerts a dominating influence, not only on how we live, but also on what we believe in. And, as books like *The Future Shock* have graphically described, the influence of science and the technology it produces is rising rapidly. So, how does science percolate into literature?

Science writing today can be broadly classified in the following categories:

1. Research papers on various scientific topics. These appear in specialist periodicals or in technical textbooks and monographs. They have limited readership, being of interest only to those who are doing research in that (limited) area.

2. Review articles and books aimed at a wider readership. These provide state-of-the-art descriptions of an overall area, with references to source material that give greater details.

3. Articles and books covering a broad area of science with readership large enough to include educated laypersons. These are 'reader friendly' in the sense that they do not demand expert knowledge of the field on the part of the reader.

4. Encyclopedias and similar works that cover, ideally all important topics listed alphabetically, with each topic given a brief description, and information on sources where further details are available.

5. Science fiction which includes stories and novels as well as poetic works with a scientific core, but which uses some interesting plot to make the

whole work readable. The idea is to use the plot to keep the reader engrossed while he or she assimilates the underlying science.

As a professional scientist I have contributed to scientific literature on all of the five fronts. So what I have to say will be derived from my personal experience (or, inexperience) supplemented by what I have learnt by reading the works of others. At the outset, of course, it is taken for granted that in more recent times literary communication also includes electronic modes like CDs, DVDs, audio discs, etc.

To begin with, the following statement may sound paradoxical but it happens to be true: *The task of the communicator becomes progressively more difficult as one proceeds down the above list.* Writing a research paper or delivering a seminar talk on it is easier than explaining a scientific truth to a layperson. For the speaker of a technical seminar (or the writer of a research paper) does not have to worry about audience comprehension: the typical listener (or reader) is assumed to be competent enough to understand what is being said. In my opinion, the above requirement of being reader-friendly is even more necessary for the writer of science fiction.

For, in the last analysis, what is science fiction? By way of definition one could lay down the rule that it is a story or novel in which some scientific principle or scientific background plays a significant role. Now imagine a reader who is *not* science friendly. Not to lose such a reader, the author may be tempted to *explain* any scientific technicalities, but this diverts the story towards a pedagogical exercise. I have seen several science fiction stories degenerate into a classroom teaching exercise: so much so that our target reader is too bored to continue. In short, the writer of science fiction has to walk on a razor's edge between (1) the scared reader who fears that the science implied by the author may be too tough to understand and (2) the bored reader who does not want classroom teaching hurled at him!

This consideration is applicable to writings of type 3 also. I recall once being asked to talk in a distinguished lecture series called the *Vasant Vyakhyaanamaalaa* (Spring Lecture Series) held in Pune. This is an annual event extending over several weeks in springtime, in which scholars are expected to talk down to an educated audience on subjects of their interest. This activity was initiated by the freedom fighter and political leader Bal Gangadhar Tilak to stimulate the cultural psyche of Indians in the British Empire. Although the unstated objective was to help Indians to be more intellectually and politically aware, the lecture series continued even when India became independent: and its success led to similar lecture series in other towns of the State of Maharashtra.

For my talk in this series I had chosen the topic of futurology, describing what one can predict about the state of the world in the next, say, 50–100 years. Since speculations about what lies ahead are always of interest to lay audiences, I was confident that I had chosen a very audience-friendly topic. However, I was in for a surprise. A few hours before my talk, I ran into a distinguished professor of Sanskrit from a local college who was known for his academic achievements. After assuring me that he was looking forward to attending my talk in the evening, he added: "But I am afraid that what you are going to say will go over my head!" This assumption on his part reflected the existing gap between "two cultures". The remark indicated that a scientist is at a disadvantage when talking to a layperson. As a 'default option' the scientist stands out amongst other speakers as a difficult speaker to understand. The professor's comment put me on special alert that evening, ensuring that nothing I said would go over the heads of my audience.

I felt rewarded that evening when the very same professor sought me out after the lecture to assure me that he had understood and enjoyed my talk.

So far as science fiction is concerned, out of the above five modes of writing, it alone is subject to the criteria for a 'good literary piece'. A piece of fiction may be excellent in bringing out the crucial role of science. Yet as a piece of literature it may fail or score low marks. I will return to this aspect, and also to literary criticism as applied to science fiction, in a later section.

Why Write Science Fiction?

I am sure different authors have different motivations for writing science fiction. In fact I have heard literary purists argue that writing with any motive degrades the quality of the writing. To me this argument does not carry much weight, as will become clear shortly.

To begin with it is worth narrating how Fred Hoyle was motivated to write science fiction. As is well known, he was one of the most imaginative scientists of the last century. In the 1950s he had written a scientific research paper in which he proposed the idea that a typical galaxy like ours has vast interstellar clouds of molecules, both inorganic and organic. In those days the emerging science of radio astronomy had yielded evidence for clouds of neutral hydrogen spread over vast interstellar spaces. There was also evidence for ionized hydrogen in regions of high temperature (where starlight had heated the gas). But the astronomy and physics communities were just not willing to admit the existence of interstellar molecules. So Hoyle's paper was rejected by both astronomy and physics journals. Nevertheless he felt that the idea was important enough to publish and the way he found to do so was to write a science fiction novel based on the idea! Thus he came to write the novel *The Black Cloud* and it became immensely popular. As a historical postscript we may add that the next decade saw the discovery of molecular clouds, just as Fred Hoyle had proposed.

The mechanism that established the existence of molecules in space used receivers of waves of around a millimeter in length. Calculations of atomic and molecular physics tell us that a typical multi-atom molecule has atoms oscillating and/or rotating. A change in the state of atoms may lead to a spontaneous transition of the molecule to a lower energy state with the result that the energy lost in the process appears as a pulse of radiation of specific wavelength, usually in the millimeter wave region. The estimates of detection and theoretical calculation are so accurate that scientists have likened the identification of the source molecule with the radiation received to the identification of a criminal with his or her fingerprint record.

By now, a vast variety of organic and inorganic molecules has been discovered in space. We even find polymers with long chainlike structures. These findings have raised expectations of another favourite science fiction idea: the existence of extraterrestrials. We will consider it later in this account.

Having given Fred Hoyle's example, I now come to my own perspective for writing science fiction. To me science fiction is a means of introducing science to the lay reader. Perhaps the example of the classic Sanskrit book the *Panchatantra* will help explain my motive! This book comprises five independent volumes, each containing a long story broken into several parts, each of which itself contains a story by way of illustrating some truth in the main theme. Sometimes even these stories also contain sub-stories with some moral. The entire book arose from the teachings of a wise scholar who was engaged by a wealthy man to tutor his sons. The father had discovered that his sons were simply not learning anything from conventional schools. How could they be taught to become responsible citizens? What type of training would keep them absorbed? This is where he was told of Vishnu Sharma. True to his reputation, this scholar could keep the willful and naughty boys absorbed in his stories, while also making them aware of what is good and evil in life. By the time these short and long stories had been drawn to completion, the boys had become learned and wise!

My perception of science fiction is likewise a series of stories which acquaint the reader with the important influence science can have on our lives. Simply studying science as a subject is no substitute, as many readers would tend to reject textbook science as a bitter pill. Indeed, a bitter pill might be rejected by a patient, although essential for health. To get round this problem pills are often sugarcoated. Likewise science fiction may be considered a sugar-coated pill which is easily swallowed. The science underneath may be made more palatable with a coating of fiction.

With the growing power that science wields, along with its technology off-shoots, it is necessary for society to know, understand and learn to cope with it. Textbook science may be a turn-off for many who might otherwise be attracted to science through the fiction format. As in the Panchatantra, stories may be such as to acquaint the reader with the subtleties of science, its power, its good and bad effects, and its ability to help in coping with natural forces. A futuristic tale may alert the reader against possible dangers in future. It may also be possible to highlight uses of a future technology through a story or a novel. In short, there is plenty of scope in this vein for a writer of science fiction.

This is my reason for writing science fiction, which, of course, may not be the motive for another writer of this genre. I would like to give some examples from my writings but before I do so, let me describe how I got into this field.

My Maiden Attempt

Having read *The Black Cloud* and some other science fiction novels I felt the urge to write a science fiction story myself. The opportunity came in 1974, two years after my return to India to join the Tata Institute of Fundamental Research (TIFR) in Mumbai. I felt that if I were to write a science fiction story, I should do so in my mother tongue Marathi. There is a vast collection of science fiction in English, mostly set in a foreign (Western) environment, and my writing yet another one would be like adding a drop to the ocean. On the other hand this genre was very rarely handled in Marathi and in a local environment, so there was a chance that the impact of a purely Marathi story would be more noticeable. While I was hesitating over this, an opportunity presented itself.

Every year, the NGO from Mumbai called Marathi Vijnan Parishad (MVP) which promotes scientific activities in the Marathi language conducts a short story competition focusing on science fiction. The assigned word limit is 2000, and the story has to be original. The top story, as judged by a panel of referees, is awarded a modest cash prize in a special ceremony during the Annual Convention of the MVP. When the call for stories for 1974 was announced, I decided to enter the competition.

I remember in October 1974, I happened to attend an astrophysics conference at the Physical Research Laboratory in Ahmedabad. While listening to a particularly uninspiring lecture, I was afraid that I might drop off to sleep, when the idea came to mind…why not start writing the story now? I had the conference writing pad with me and already had some ideas on the plot of the story. As I wrote, the words came naturally and there were very few pauses. The speaker, in case he looked in my direction, might have been flattered by the (mistaken) impression that I was feverishly taking down notes from his lecture! Anyway, nearly a third of the story was written up in that initial spurt of writing.

In due course, I completed it and entitled it "*Krishna Vivar*" (Black hole). The idea, let alone the name, 'black hole' had not yet caught on in India. I used the concept of clocks going slow near the black hole horizon. This being my first attempt at story writing, I was not confident of success. However, I needed to take two precautions when submitting my story to the MVP.

Those were the days before word processors became common. Even typewriters using Marathi were not very common, Thus, the entries for such competitions were usually in the handwriting of the authors. It was likely that my handwriting would be recognized by the MVP officials since on several occasions in the past I had corresponded with the organization. So I not only adopted a different name but also submitted the story in my wife's (Mangala's) handwriting. I chose the fictitious name "Narayan Vinayak Jagtap" whose initials NVJ were the reverse of mine, JVN. The correspondence was carried out under the address of the TIFR.

Thus I submitted my story giving no hint of my real identity, so as to avoid any possibility of the MVP figuring out Jagtap's real identity.

In due course the results were announced and it so happened that the science fiction story "Krishna Vivar" was declared the best story. As its author, Narayan Vinayak Jagtap was invited to attend the forthcoming annual convention of the MVP and receive the prize of a hundred rupees from the hands of the President of the Parishad.

It was then that I chose to reveal Jagtap's true identity, while requesting the award money to be given to some charity. My clandestine attempt for the prize was greatly appreciated by the MVP as well as the intelligentsia in Maharashtra (the state whose official language is Marathi). Many people felt that my participation would raise the status of the competition, encouraging many more to participate in the future events.

But that was not the end of the story! Shortly afterwards the annual gathering of literary big shots in Marathi took place in the town of Karad, and in her opening address, the President Durgabai Bhagvat, a distinguished author and critic, made a reference to my story, welcoming its appearance with the hope that the genre of science fiction would bring an additional dimension to Marathi literature. Known for her very critical views, this was high praise indeed and boosted my morale no end! Many literary experts felt that Durgabai's endorsement would promote the popularity of science fiction, not only with readers but also with writers.

For me one positive aspect was that one famous literary journal of Maharashtra, called *Kirloskar*, invited me to contribute science fiction stories as and when I wrote them. The editor Mukundrao Kirloskar commanded great respect for his literary acumen. Certainly his backing was a great tonic to the budding literary genre. Thanks to his encouragement I felt courageous enough to venture further into the field of science fiction.

My first story published by *Kirloskar* was on Ganesha with the trunk turned to the right. Ganesha is the name given to an idol with the head of an elephant on a human body. As usual, Hindu mythology had an interesting tale about the genesis of this combination. Initially, Ganesha had an all human body created by Parvati, the goddess wife of Shiva. Parvati appointed her creation as a guard outside her cave wherein she was enjoying a bath under a natural waterfall. Ganesha's duty was to stop any intruder from disturbing his mother while she was having her bath. Her orders did not single out any exception to this rule, and they therefore applied to Shiva himself. Perhaps he was not expected during the bath time. But he did show up and, as per his mother's orders, Ganesha stopped Shiva from entering. Shiva did not know that this boy had been created by Parvati and was only doing his duty according to her orders. Likewise Ganesha did not know that Shiva was in *loco parentis* to him. This lack of information led to a fight in which Shiva beheaded Ganesha. At this point Parvati came out, having finished her bath. She was aghast at what Shiva had done and Shiva too was contrite having learnt the story behind Ganesha's creation. To pacify Parvati he promised to revive the dead boy. But his head could not be found and the only substitute Shiva could lay his hands on was an elephant's head. He promptly set it on the body of Ganesha and revived him.

This was the story of how Ganesha came into being. All the lore associated with him has endeared him to the masses and he is the most popular deity to be approached for success when setting out on a new venture, or again when seeking help and protection for some family function. Typically his idol shows him in a benign mood with his trunk turned to the left, that is, on the side of the heart. There are rare idols where his trunk is turned to the right. Usually, but not always, this version shows Ganesha in an angry or aggressive mood, while an idol with a benign mood but trunk to the right is supposed to be rare. I have described this background in some detail here, in case a reader unfamiliar with Hindu mythology reads the story.

Likewise, for the reader unfamiliar with the notion of symmetry in fundamental physics I may add the following note. Suppose we observe a phenomenon in a room, like the dropping of a plate from a table. If there is a mirror in the room placed so that we can see the phenomenon reflected in it, the mirror image will show the same phenomenon. However, if we similarly watch a book fall from the table, we will see the same phenomenon but with a difference. The title of the book seen on the front cover will be reflected in the mirror view. In an English title, letters like A, H, M, etc., are unchanged while letters like B, C, N, etc will appear in an unfamiliar form. At the sub-atomic level, in general, particles like the electron and the proton have mirror counterparts which are also found in nature. Had this been true for all such particles, we would have said that they form a distribution which is left-right symmetric. However, there are particles called neutrinos that are always found to spin in an anticlockwise direction. Their mirror counterparts spinning in a clockwise direction are *never* found in nature. This result is often stated by saying that "the universe is not left-right symmetric." Or in a more dramatic form that "God is left-handed".

But, if we are dealing with macroscopic objects like the idol of Ganesha, we will find that its mirror counterpart can exist in nature and the story describes how one can be transformed into the other. One uses the idea of a Mobius strip which effectively transforms a flat (two dimensional) figure into its mirror counterpart. Can the same be done in higher dimensions? In a popular science book the scientist George Gamow argued that this could happen if the space in the universe had a twist! A left shoe sent round such a space would come back as a right shoe. The idea is intriguing and there are extensions of Einstein's general relativity in which swarms of coherently spinning particles can produce twists in space. The story tells the consequences of having a machine that produces such a twist in a limited region. An object going round a region like this would be transformed into its mirror image!

I will not go into the details of the story, but will just say that the plot was appreciated by Indian readers.

Vaman and Other Novels

Perhaps this is the right place to recall my second science fiction novel *The Return of Vaman*. It is generally assumed that a science fiction story must have an extra-terrestrial character. Nothing can be farther from the truth. H.G. Wells wrote about the *Invisible man* in a purely terrestrial context. Jules Verne in *Round the World in Eighty Days* or in *Twenty Thousand Leagues Under the Sea* confined himself to the terra firma. The extra-terrestrial setting with aliens and their strange devices gives a greater scope to the writer's imagination; but the more limited setting on the Earth has the advantage that the reader can more readily identify himself with what is going on in the story.

It was partly with that view that I wrote the *Return of Vaman*, my second science fiction novel. The trigger for this novel was an underground gravity experiment that had been set up in Gauribidnur near Bangalore. The description of the plot is again superfluous for the reader who has read the novel already, or a spoil-fun if he has not. But again, for those new to Hindu mythology, an explanation of the origin of the name Vaman may be necessary. Vaman was one of the many incarnations of Vishnu (the God who protects the universe). He was created to rid the Earth of a demon King Bali who ruled the underground world of *Patal*. The reader needs to know that the surface of the Earth accommodating man and other living beings had the heavens (abode of the various gods, ruled by King Indra) on top and the Patal (abode of demons, ruled by King Bali) below. Although Bali was well behaved, he was seen as a threat to the two upper tiers and a plot was hatched to get rid of him. Thus Vaman, a dwarf, was deputed to the mission.

Bali had been giving away a lot of his wealth in a holy sacrifice and Vaman turned up to seek alms. Since he never turned away anybody who wanted alms, Bali asked Vaman what he could do for him. Vaman asked for just the space covered by three of his paces. How much ground would a dwarf thus cover, thought Bali, and readily agreed to donate that much land. Thereupon Vaman grew and grew until three of his paces covered all that Bali possessed, and finally he was buried underground by Vaman's final step. In short, what was seen as a diminutive and charming figure turned itself into a very dangerous threat to Bali and his kingdom.

This mythological history may help the reader appreciate why the name of Vaman comes into the science fiction novel. The robotic Vaman starts off as a loveable figure, always ready to help. But would he continue that way? This question is answered in the final denouement of the novel.

My first novel was *The Message from Aristarchus*, with a story based on the SETI programme. The plot centres around a radio astronomer who manages to send out a coded message in a signal from a telescope strictly dedicated to the defense programme. His action leads to a whole range of consequences, including aliens acting in response. The novel runs through its menu of messaging, romance and suspense and ends on a question mark designed to haunt the reader about what the future might bring.

Virus was my third novel. Although based on the Earth, it has an extraterrestrial component which becomes apparent in the final climax. It was inspired by the Giant Metrewave Radio Telescope (GMRT) which was under construction near Pune at the time it was being written. It describes scientists and their politics against the background of the large telescope. Again, it would be a give-away to describe how the novel ends.

During the 1990s I was approached by the Sahitya Akademi (the leading national literary body) to write a novelette for teenagers. I expanded my short story into a novel entitled *A Cosmic Explosion*, which I refer to later in this essay. Around the year 2000–2001, I wrote a novel in Marathi which I hope to translate into English myself. In Marathi it is called *Abhayaranya*, meaning 'sanctuary'. The basic idea here is that our Earth-based life is under scrutiny by advanced and benign extraterrestrials who treat it as a sanctuary for evolving life. Occasionally, they may interfere at some crucial juncture, usually guiding an exceptionally imaginative human being to some new discovery that helps in the evolutionary progress of the civilization. They might also intervene more drastically if we on this planet contrived to put ourselves under a threat of total annihilation.

I will now move on to my method for appraising a science fiction story… in short, I shall describe what I call *good* science fiction and what I single out as *bad* science fiction.

Good Science Fiction

Before coming to my way of assessing whether a particular science fiction story is good or bad, I must recall a memorable episode in 1964. I was then visiting Caltech for a semester and Fred Hoyle, my research guide, was also spending some time there. While I recall that visit for various important happenings, the one I need to emphasize here relates to a debate that took place in Caltech's Beckmann Auditorium.

That evening the Beckmann was overflowing with student and faculty. Hardly surprising, because the occasion was a debate on the topic: *The Message of Science Fiction: Prophetic or Profane?* The debaters were two provocative minds: Fred Hoyle who occasionally wrote science fiction stories or novels but was a distinguished astrophysicist, and Ray Bradbury, a very well known author of science fiction. While these scholars examined the existing literature for good or bad science fiction, perhaps the most pertinent point was made by Ray Bradbury. He observed that, for a person like himself who was born around the time of World War I, the four to five decades of his life were littered with cases of science fiction becoming reality. For, at the time of his birth, the concept of an atom bomb existed only in imagination, and so far as sending man to the Moon was concerned, there had been no real expeditions, only the science fiction of Jules Verne. And there were plenty of other ideas that would later become realities, like computers, antibiotics, guided missiles, and the list goes on.

In short, many ideas that were considered esoteric and way ahead of reality were actually realized during those four to five decades. Perhaps here is a clue that tells us what "good" science fiction is like: it anticipates the advances of science. Despite the gap between perception of future science and reality, good science fiction anticipates the former. It has been commented by many, for example, that Jules Verne's description of travel to the Moon comes uncannily close to what actually happened with Apollo 11.

The ability to predict what will happen in science and technology in the years to come may well be an attribute of good science fiction. In this role the writing may predict some harmful side-effects of a technology already existing today. Or, on a more positive note, it may indicate a possible future technology that will help solve today's problem.

In a technical article written in 1945, the distinguished science fiction writer Arthur C. Clarke wrote about the possibility of geostationary satellites. These are satellites orbiting the Earth at a height of around 40,000 km, where their orbital period matches the Earth's period of rotation around its North-South axis. Such a satellite will appear stationary when viewed from a fixed point on the Earth's surface and can be used to provide information services like fax, email, TV programmes, etc. Such satellites can also conduct country-wide classroom programmes that provide education to remote or inaccessible locations.

I am not sure that all science fiction writers would share one of my reasons for using this genre: that is, to promote the way science can benefit society. As this is meant to be a personal account, I would like to share some examples of my stories with the reader. In *The Comet*, the main character is Dutta, a grandfatherly figure who happens to be an amateur astronomer. His long-standing wish to discover a new comet is finally realized when he does indeed find one and it is named after him. However, *Comet Dutta* turns out to be heading dangerously close to the Earth. When scientists discover this likely catastrophe, they hold meetings to carry out detailed studies of the comet, finally reconfirming the likelihood of a collision with the Earth. Can it be avoided? The scientists debate, discuss and finally converge upon a strategy to prevent the collision. Although the discoverer Dutta himself had not progressed beyond high school mathematics, he was appreciative of what the scientists were trying to do. The scientists also grew attached to this unassuming but highly practical person and kept him a part of their secret mission to waylay the comet.

The mission had to be secret because of the likely panic that would spread rapidly if it were known that a comet was to strike the Earth. In India, known for many old superstitions, there was already unease at the knowledge that an Indian had been the discoverer of the Comet Dutta. Indeed his family, relations and neighbours wanted Dutta to conduct some yajna (holy sacrifice) to counter the ill effects the comet would bring. This he flatly refused to do, himself being a wholly rational human being. And he was aghast to discover that, given his refusal, his grandson was made to conduct all the rituals! The priests had argued that, instead of Dutta, any descendent of his would do for the rituals to be conducted. Dutta's wife was a traditionalist who was easily carried along by such arguments.

The scientists had meanwhile constructed and launched a space probe carrying gas and a nuclear explosion device which was heading for the comet. It was to be exploded in a cloud of gas close to the comet, thus giving it a small push sufficient to change its direction. If the plan succeeded, the push would be enough to ensure that the comet went close to the Earth but passed by at a safe distance.

Dutta could not help contrasting the two attitudes: one purely superstitious and the other rational. When the mission against the comet succeeded he thought that he could now reveal the real situation and explain to his wife why the Earth had remained safe, despite the threat brought by the comet. But, alas, she argued that the trick was to be imputed to the yajna conducted on his behalf by his grandson. This was where the story ended, highlighting the contrast between the superstitious and the rational. For me, however, it was a matter of some satisfaction that I had got the scientific idea right, when at a press conference about a decade later, a NASA scientist advocated the very same method for diverting a solar system body heading for collision with the Earth.

Collisions within the solar system are not all that common, but they have happened. In 1994, the comet Shoemaker-Levy crashed into Jupiter and this outcome was not only well-documented, but had also been predicted to happen down to the day and date. Jupiter is somewhat exceptional in the sense that it has large enough mass to attract gravitationally an object that happens to come close enough. Thus it will *increase* the probability of collision. The Earth in comparison is much less massive and so less prone to such collisions. Even so it also bears the marks of such impacts in the form of craters with diameters measurable in kilometers. The Arizona Crater in the south-west of the USA is a well visited tourist attraction. A bigger and more attractive spot is Lonar in the Buldana district of the state of Maharashtra in India. It has the form of a crater lake. It was earlier mistaken for a volcanic crater, but now it is believed to have been caused by the impact of a twenty million ton stone or *meteor* which hit the Earth nearly 50,000 years ago. The impact is calculated to have generated an energy equivalent to nearly five hundred atomic bombs, each comparable to the one that destroyed Hiroshima. Although the energy released was not nuclear but thermal, the rise in temperature melted the stone meteor and it got mixed with underground water. This changed the mineral composition of the latter, which is why the chemical composition of Lonar Lake is considerably richer than that of a typical water reservoir in the neighbourhood. And of course the combustion started by the release of so much heat led to consumption of atmospheric oxygen, thus destroying life over a region much larger than the impact site.

For this reason astronomers felt that there should be an archive of collisions in the solar system, including not just the "have beens" but also the "likely future impacts". So a programme called "Operation Sky Watch" was initiated in the United States with the aim of computing the future orbits of all solar system bodies to check the when and where of future collisions. This database will be able to warn future Earthlings of any likely threats of this kind.

This story has two morals! The main theme tells us how an international team of scientists, working under the aegis of the UN, plans and successfully completes the job of saving our present civilization from the destructive collision of a comet. However, a more worrying aspect is presented by the side-show of rituals conducted in the comet discoverer's house to ward off any evil effects of the comet. Although the discoverer himself is rational, his family and friends are bound by superstitions. My underlying purpose in this story was to show the contrast between the two extremes that prevail in our country today.

I will end this section with an idea cleverly used by A.A. Milne, the celebrated author of *Winnie-the-Pooh*, in a fairy tale. The prince was hoping to free a princess trapped in a witch's castle. A friendly fairy gave him special shoes with the help of which he could take paces as long as six miles. Unfortunately, the witch's castle was only three miles away. So how could he reach the princess with his fast shoes? The solution lay in drawing an isosceles triangle each of whose long sides measured 6 miles while the base was three miles. This is a simple idea involving school geometry. It shows how one can construct any length span not longer than the range of the shoes supplied by the fairy.

Indeed, one can summarize the above discussion by highlighting this aspect of good science fiction: it displays the power of a scientific idea.

Bad Science Fiction

Good science fiction can be socially relevant, sometimes generating ideas that may be helpful to society, and it may be able to add to existing science, besides being thrilling and mind-stretching. Unfortunately, the fraction of good science fiction is small and it may be hard to spot the good ones among the vast majority of bad science fiction.

The bad variety can be of various kinds. Often, after the thin vaneer of 'science' is removed, what shows up is a fantasy or a horror story. Now I have no objection to horror stories *per se*: I have enjoyed reading a few like those by Bram Stoker, Conan Doyle, Alfred Hitchcock, and others. These are plain horror stories. What I object to are those that advertise themselves as science fiction and are no different from tales of horror. If one is expecting science fiction to make the lay reader science-friendly, then these horror books will not serve that purpose. Rather they will create or foster a distorted image of science in the reader's mind.

Films like the *Star Wars* series form another class of what I call pseudo-science fiction. With a simple mental exercise one can transform them into classic "Westerns" with horses and wagons replacing weird creatures and spaceships. The long series "Star Trek" was more imaginative and closer to science, but it suffered from another problem which I will highlight next.

Einstein's special theory of relativity has mystified a lot of people who generally only remember, let alone understand, two basic aspects of the theory. One is the maxim that you cannot travel or send information faster than light. The other is the equation $E = Mc^2$, equating mass with energy. Most science fiction stories in the bad category simply ignore the light speed limit. They have spaceships travelling across our Milky Way in times of the order of a few years when light itself would take 100,000 years! I feel very uncomfortable when such liberties are taken in the course of a science fiction story. Break this rule and then you do not worry if the spaceship *Enterprise* travels great distances in the Galaxy within relatively short time spans. But I wish the author would describe how his heros (or villains!) managed to break Einstein's law.

As I mentioned earlier, Jules Verne's novel *Round the world in eighty days* reveals a scientific truth only at the end, thus making the climax particularly effective. This is an example of *good* science fiction. By contrast, bad science fiction would either have nothing scientific to offer or, if it does offer some maxim, it will turn out to be wrong!

Finally, bad science fiction often ignores any limits of time and space, tolerating ranges where causality fails…and even situations where effects precede causes. While scientists discover more and more about genetics and cloning, poor science fiction seems to go on applying its own biological laws. Of course, as I stated earlier, a science fiction story may involve new laws of science, but it should manifestly avoid conflict with well tested science that has already been established.

Some Examples of Logical Constraints I Have Respected

I give two examples of the logical restrictions I have had to face while writing. In the story *A Cosmic Explosion*, I wished to demonstrate the harm a supernova could cause. Of course, the most spectacular supernova explosion is associated with the Crab Nebula, which was observed on Earth on July 4, 1054 AD. We know the date because of the records kept by the Chinese and Japanese astronomers of the time. That supernova lies more than 6000 light years away. That is, what we see there today happened more than 6000 years ago. Any harmful ingredient from that explosion, like cosmic rays, would take more than 6000 years to travel to us. Cosmic rays may suffer some resistance in travel because of magnetic fields or matter they encounter en route, so they are not expected to travel as fast as light. Assuming that cosmic ray particles travel at an average of two thirds of the speed of light, they would arrive take, say, 9000 years to arrive here. That is, they would reach here three thousand years after the Chinese saw it and hence two thousand years from today. This has to be borne in mind when we investigate whether the Crab explosion will be harmful to us. For the distance is so large that it is too early to expect the debris from that event to arrive in the near future, or indeed to expect its effect to be large.

Keeping these caveats in mind while drawing up the plot for my story, I decided to divide the time span into three periods and made the supernova explosion considerably closer than the Crab event. So it was in Period I that I had the explosion observed in around the seventh century, at a time when India was enjoying a relatively prosperous spell. That of course meant that the explosion was actually much earlier. A Buddhist monk of considerable knowledge and wisdom arranged to have records of the event kept underground, safe for posterity. Keeping track of time in this way, it was in Period II, in 1996, when the records of the observation were dug out and read by archaeologists. These showed the way towards further investigations. It quickly became clear that cosmic ray showers would arrive sooner rather than later, and that their effect might be fatal. Finally, Period III concerned the year 2710 when the Earth was slowly recovering from that catastrophe.

By keeping to these time constraints, it becomes clear to the reader that, while a terrestrial explosion is a short term phenomenon, a cosmic one could be an event lasting two millennia!

My second example is drawn from the novel *A Message from Aristarchus*. It centres round radio messages sent to a likely civilization. While such messages travel with the speed of light, the receiving location cannot be too far away. For example, if it were even just 100 light years away, it would take at least 250 years and probably more for a message to be received by the aliens and for them to actually arrive on Earth. Thus a novel requiring human-alien interaction with the present level of technology will have to have aliens not more than 10–20 light years away. This and the earlier example both alert the author of a science fiction to be aware of these space-time restrictions.

My Views on Some Science Fiction Films

How have the film and TV media received science fiction? In the early 1960s the BBC aired a television serial called *A for Andromeda*, with a scheduled length of 13 episodes. The principal author was Fred Hoyle, while John Eliot co-authored the script. The serial was well received, so much so that it was followed by a sequel comprising another 13 episodes with the title *Andromeda Breakthrough*. The role of the heroine in the first serial was played by Julie Christie, a more or less 'raw' recruit from her film and TV course. This role very successfully projected her as an excellent actor. Ironically, this success raised her 'level' in professional circles so much that, when she was approached to play the same role in the sequel, the BBC discovered that she had been 'priced out' of their budget! So another girl had to play that role.

The success of these serials raised the fraction of science fiction programmes on TV. A very long running programme was *Dr Who*, which was shown on the BBC in the early evening. With interruptions, the series has so far had 812 episodes up until November 2014, and may probably claim some kind of world record. The earlier version of this serial had 25-minute episodes each connected to the next, while the later (revived) version had 45-minute programmes, each more or less standing on its own. The Guinness Book of World Records mentions another serial entitled *Smallville*, with a run of 218 episodes broadcast *without interruption*. It could still be argued that, compared to the never-ending soap operas on TV, these science fiction series provide an intellectually superior viewing diet.

However, if one wants to see the maximum influence of special effects, one should look for full scale movies on the big screen rather than short episodes on small screens. Films like *Star Wars* and its sequels show how special effects can dominate a film and its science fiction plot. A good combination of science fiction and special effects is found in movies like *2001: A Space Odyssey*. There are mystical overtones in the original Arthur C. Clark novel. At the beginning, in his caveman guise, man gradually picks up inspiration and hauls himself up the ladder of 'progress'. The mystical part is in the form of a huge monolith with sides in the ratio $1^3: 2^3:3^3$. It emits a haunting tune which has an impact on the human brain. This tune, by the way, was borrowed by the BBC as the signature tune for its televised broadcasts of the Apollo 11 expedition to the Moon.

By comparison the movie *Zardoz* contains more fantasy, its main emphasis being on immortality. Set in the year 2293, it describes a small but elite group of people who have achieved near immortality, living in an isolated 'vortex' away from the rest of the population, who are referred to as 'brutals'. The elite have almost secured immortality, because the key to their death has been hidden in a tabernacle with safeguards preventing access. The brutals live under the heel of the god Zardoz, at whose bidding the 'chosen few' can exterminate the rest. In a sense Zardoz is like the *Wizard of Oz*, who was a humbug with no special powers. He is in fact a member of the elite group of the vortex. One of the brutals, Zed, has indeed found this out. He kills the god and comes to the vortex, where he discovers the secret of death in the tabernacle and releases the elite colony from the pains of immortality. At the end of the film, one is inclined to ponder whether we will ever find ourselves in such an elite state in some distant future, when science has somehow eliminated all the trials and tribulations of life.

In this connection, I should mention that I recently saw the science fiction movie *Interstellar*. Its theme and the technology behind its execution raised great expectations in my mind. Alas, I was to be disappointed! The science behind it is vague at its best and mystical at its worst. The use of the wormhole concept is there, but the overall role of gravity is hard to fathom. So is the environment in an alien land, depicted as not very different from where we live. Certainly the technology for special effects is intelligently used, but to what avail? What new science is being projected? To a lay viewer it may all look fantastic, but not in a science-friendly way.

The Andromeda Strain is another type of science fiction movie that combines suspense with science, without any great emphasis on special effects. A returning space probe brings in an alien life form, hostile to Earthlings. This hostility is demonstrated when, apart from two people, it kills all 66 of the other inhabitants of a Mexican village. A number of questions need urgent answers. How was death brought about on such a large scale? And why were two of the total population of 68, an old man and a six month old baby, unaffected? Would the killer source multiply and spread? Could it be controlled?

To find answers a crack team of super-scientists is engaged to work in an underground super-lab. Glimpses into their work and character are given, as are examples of domineering politicians. With all this super-atmosphere, it is something of an anticlimax to discover that, if the lab needs to be saved, the one person holding the key to the device capable of stopping all destruction only has to climb a few metres up a vertical ladder and find the appropriate keyhole. So, in the last analysis it is man's ape-ancestry that comes in most helpful by enabling him to clamber so far up!

Obviously, when a science fiction book is made into a film, there could be crucial differences of interpretation between the original author and the director and/or script writer. How they are reconciled will have a bearing on the final version of the film.

Critics and Reviewers

To end, I have a few reflections on the role of reviewers. In my language, Marathi, there are hardly any professional experts capable of reviewing science fiction novels and short stories. This, in spite of there being literary reviewers capable of doing this job for literature in general. The reason is that many literary experts are afraid of the technical envelope around the sciences. Starting out with the premise that they will not understand the science on which the fiction rests, they are unable to write a reasonable literary criticism. Some do a very superficial job, while others may stick their necks out and write something that is manifestly wrong.

I can cite a personal example. My story *The Comet*, described earlier, referred somewhere to a date, *The First of October*, which played a critical role in the scientific enterprise that underpinned the story. A critic writing a review of the book blandly stated that my story was plagiarized from Fred Hoyle's novel *October the First is Too Late*. Except for the date being the same, the two writings actually had nothing else in common. On closer questioning, the critic confessed that he had not read either work and had simply been struck by the coincidence of dates! Another scholar who got a Ph.D. on his critical review of science fiction by Marathi authors asserted that one of my stories was a copy of Conan Doyle's story *The Case of the Opel Tiara*. In fact there is no such story by Conan Doyle!

I think a good case can be made for organizing a course on how to review science fiction stories or novels. Those who offer languages for study will stand to gain from it. In Marathi especially, where the field of science fiction is rather limited, some training on how to judge a science fiction novel would certainly help to raise the overall standard.

Concluding Remarks

This brings to a close my own perception of science fiction. When one looks at the Indian national scene, one finds that Bengali and Marathi are the two leading languages fostering this genre. In Bengali in particular, the famous movie director Satyajit Ray wrote good science fiction. Science fiction for cinema or TV is even less common. In the 1980s The Children's Film Society of India produced a film version of my story *The Comet*. Although preliminary studies of some of my short stories were carried out by movie producers, they got bogged down by the challenges of depicting science and the employment of special effects.

I have on various public occasions tried to get literary figures interested in writing science fiction. But to no avail! 'Science' continues to be a scare word, and school teaching which sets science apart from humanities is partly responsible for creating a dichotomy. This creates a false impression that is thus perpetuated, according to which those good in science will be poor at humanities and vice versa, a situation that makes it unlikely that we will ever generate the ideal science fiction writer!

Nevertheless, I continue to call upon those with literary flair to take up writing science fiction, using a friend from the sciences to advise them on scientific issues. This may be one way of attaining the goal set by Literary Society President Durgabai Bhagvat.

Printed in the United States
By Bookmasters